A Guide to Signals and Systems in Continuous
Time

Stéphane Lafortune

A Guide to Signals and Systems in Continuous Time

 Springer

Stéphane Lafortune
Department of Electrical Engineering and
Computer Science
The University of Michigan
Ann Arbor, MI, USA

ISBN 978-3-030-93029-5   ISBN 978-3-030-93027-1   (eBook)
https://doi.org/10.1007/978-3-030-93027-1

This Springer imprint is published by the registered company Springer Nature Switzerland AG
The registered company address is: Gewerbestrasse 11, 6330 Cham, Switzerland

This book is dedicated to all the students who have taken my signals and systems classes at the University of Michigan.

Preface

Why another *signals and systems* book, when there are so many excellent textbooks available covering this standard material for undergraduate engineering students? The answer is that this short book, or "guide," is not meant to serve as a primary textbook, but rather as supplementary material in the context of a first course in signals and systems in undergraduate engineering curricula. My objective is to cover, in a concise manner, the main material that students need to know for continuous-time signals and systems, with a focus on linear time-invariant systems and their response to different classes of input signals. This book was built from an extensive set of slides that I developed over the years to complement blackboard/whiteboard/tablet writing when teaching this material. As such, there are few examples or figures, since these were done on the blackboard/whiteboard/tablet during lecture. It seemed worthwhile to use these slides as the basis for a short book that could be used both as a detailed summary of the key concepts and results and also as a means of reinforcing the students' understanding of the material. A significant amount of rewriting and many additions were necessary to convert the slides into a book, but the main content and organization were maintained. Since my goal was to produce a book that can serve as a handy supplement to regular teaching materials, I limited the total length to around 100 pages.

Our introductory signals and systems course in the EECS Department at the University of Michigan has evolved from a junior-level required course (previous course EECS 316) for all electrical engineering majors to a sophomore-level course, with a hardware laboratory, required for both electrical and computer engineering majors (current course EECS 216). This course covers exclusively the continuous-time case, while discrete-time signals and systems and digital signal processing are covered in a follow-on junior-level course (EECS 351). This book follows the same progression as in our introductory course: time-domain analysis of continuous-time linear time-invariant systems using the convolution integral; consideration of periodic signals and introduction of the Fourier series representation of the input and output signals; generalization to aperiodic eternal signals and frequency-domain analysis using the Fourier transform; application of the preceding technique to study filtering and communication systems, along with a brief discussion of sampling; further generalization to the *s*-domain for right-sided signals and causal systems using the unilateral Laplace transform, along with complete solution of linear constant-coefficient differential equations with non-zero initial conditions; and, finally, a brief introduction to feedback control and PID compensators.

I hope that this book will be a useful guide to undergraduate engineering students who want to learn about linear time-invariant systems and specifically *why*, *when*, and *how* to use transform techniques à la Fourier and Laplace in their study.

References

The material in this book is standard and can be found in signals and systems textbooks. I have not tried to attribute original citations to the results herein, since many of them are centuries old. Many existing textbooks contain worthwhile historical remarks and can be consulted in that regard.

My principal source of inspiration in the writing of the course slides upon which this book is based was the course notes prepared by my colleague Kim Winick (now professor emeritus) when he developed the syllabus of EECS 316 at the University of Michigan. Two other main references I used heavily are the textbooks by A. Oppenheim and A. Willsky (with I. Young) [6] and by S. Soliman and R. Srinath [9]. Over the years, and during the final writing of the book, I also consulted several additional textbooks that we used as textbooks or suggested references in EECS 316, 306, and 216:[1] Senior [8]; Phillips and Parr [7]; Haykin and Van Veen [1]; Lee and Varaiya [5]; Kamen and Heck [2]; Kudeki and Munson [3]; Ulaby and Yagle [10]; and Lathi and Green [4]. The biographical information that appears in several footnotes was obtained from Wikipedia.

Acknowledgments

I wish to thank my colleague Kim Winick, whose teaching materials provided the foundations for this book. His vision on what material to teach, and in what order, in the first course in signals and systems at the undergraduate level is reflected in this book. It is a pleasure to also acknowledge the other colleagues with whom I have taught this material over the years: Achilleas Anastasopoulos, Jessy Grizzle, Vijay Subramanian, Gregory Wakefield, and Andrew Yagle. I have learned from all of them. Finally, I thank Aaditya Hambarde for his careful reading of the manuscript. Of course, all errors remain my responsibility.

Ann Arbor, MI, USA   Stéphane Lafortune

---

[1] I refer to the editions that I used at the time, which may not be the most current ones.

Contents

1 Introduction 1
   1.1 What Does "Signals and Systems" Mean? 1
       1.1.1 What Is a Signal? 1
       1.1.2 What Is a System? 2
   1.2 Preview of This Book 4
   1.3 Properties of Signals 6
       1.3.1 Symmetry and Periodicity................................. 6
       1.3.2 Notions of Energy and Power Signals 7
   1.4 Signal Transformations.. 8
       1.4.1 Combinations of Time Transformations.................... 9

2 Systems 11
   2.1 Properties of Systems 11
       2.1.1 Memorylessness and Causality 11
       2.1.2 Zero-State Response and Zero-Input Response 13
       2.1.3 Stability 14
       2.1.4 Time-Invariance and Linearity 15
       2.1.5 Complete Response of LTI Systems 16
   2.2 LTI Systems: Time-Domain Analysis 17
       2.2.1 Unit Impulse Signal 17
       2.2.2 Impulse Response... 20
       2.2.3 Convolution Integral Theorem 20
       2.2.4 Properties of Convolution 22
       2.2.5 Graphical Convolution 23
       2.2.6 System Properties and the Impulse Response 25
   2.3 LTI Systems: Frequency-Domain Concepts......................... 25
       2.3.1 Eigenfunctions of LTI Systems 26
       2.3.2 Frequency Response Function and Transfer Function 27
       2.3.3 Response to Sinusoidal Inputs: Sine-In Sine-Out Law 30

| | 2.3.4 | Response to Sums of Sinusoids | 32 |
| | 2.3.5 | Complex Impedances for Circuits and Second-Order Systems | 32 |

3 Periodic Signals and Fourier Series 35
   3.1 Fourier Series of Periodic Signals 35
      3.1.1 Exponential Form 35
      3.1.2 Combined Trigonometric Form 37
      3.1.3 Trigonometric Form 37
      3.1.4 Calculation of Fourier Series Coefficients 38
      3.1.5 Parseval's Theorem 40
      3.1.6 Interpretation and Convergence Issues 41
   3.2 Fourier Series and LTI Systems 42

4 Analysis of Stable Systems Using the Fourier Transform 45
   4.1 The Fourier Transform of Continuous-Time Signals 45
      4.1.1 Derivation and Existence Conditions 45
      4.1.2 Linearity, Frequency Spectra, and Bode Plots 48
      4.1.3 Transforms of Common Energy Signals 49
      4.1.4 Transforms and Inverse Transforms of Impulses 50
      4.1.5 Transforms of Some Power Signals and of Periodic Signals 50
   4.2 Properties of the Fourier Transform and Their Applications 52
      4.2.1 Convolution and Modulation Properties 52
      4.2.2 Other Properties 53
      4.2.3 Parseval's Theorem for Energy Signals 56
   4.3 Filtering 57
      4.3.1 Ideal Filters 58
      4.3.2 DSB-SC AM and Synchronous Demodulation 59
      4.3.3 Frequency-Division Multiplexing 60
      4.3.4 Realizable Filters 61
      4.3.5 Notions of Bandwidth of Signals and Filters 62
   4.4 Key Takeaways 64

5 Sampling and Reconstruction 65
   5.1 Sampling Theorem 65
   5.2 Ideal Sampling and Reconstruction 66
   5.3 Practical Sampling and Reconstruction 69

6 Analysis and Control of Systems Using the Laplace Transform 73
   6.1 The Laplace Transform: What Is It and Why Do We Need It? 73
      6.1.1 The Two Types of Laplace Transforms 74
      6.1.2 Two Key Results 75
      6.1.3 Why Use the Laplace Transform? 76
      6.1.4 Region of Convergence 77
      6.1.5 Some Important Transforms 78

6.2   Inverse Laplace Transform 78
     6.2.1   Partial Fraction Expansion 80
6.3   Properties of the Laplace Transform 84
6.4   Solving Differential Equations Using the Unilateral
     Laplace Transform 85
     6.4.1   Transfer Function (Revisited) and ZSR 85
     6.4.2   ZSR and ZIR 86
6.5   Pole Locations, Stability, and Time Response 88
     6.5.1   Pole Locations and Stability 89
     6.5.2   Summary 91
     6.5.3   Oscillations in Time Response 92
     6.5.4   Final Value Theorem 92
     6.5.5   Time Response of Second-Order Systems.................... 92
6.6   Feedback Control: A Brief Introduction............................. 93
     6.6.1   Proportional Feedback....................................... 94
     6.6.2   PID Compensators... 95
     6.6.3   Pole Placement Using PD Compensator..................... 96
     6.6.4   Zero Steady-State Tracking Error Using PI Compensator... 96
6.7   Key Takeaways 97

A   **Common Signals** 99
   A.1   Continuous Signals 99
   A.2   Piecewise-Continuous Signals 100

B   **Proofs** 103
   B.1   Proofs of Main Results 103
     B.1.1   Convolution Integral Theorem 103
     B.1.2   BIBO Stability and IR 104
     B.1.3   Parseval's Theorem 105
   B.2   Proofs of Fourier Transform Properties 106
     B.2.1   Convolution and Modulation................................ 106
     B.2.2   Other Properties... 107
   B.3   Proofs of Laplace Transform Properties.............................. 108
     B.3.1   Convolution and Time Differentiation 108
     B.3.2   Time Integration and Time Shifting 109
     B.3.3   Final Value Theorem 110

References.. 111

Index.. 113

Acronyms

ADC	Analog-to-Digital Converter
AM	Amplitude Modulation
BW	BandWidth
CLTF	Closed-Loop Transfer Function
CLTF-C	Closed-Loop Transfer Function with Compensator
CLTF-P	Closed-Loop Transfer Function with Proportional compensator
CLTF-PD	Closed-Loop Transfer Function with Proportional-Derivative compensator
CLTF-PID	Closed-Loop Transfer Function with Proportional-Integral-Derivative compensator
CT	Continuous Time
DAC	Digital-to-Analog Converter
DSB-SC	Double-SideBand Suppressed-Carrier (Amplitude Modulation)
DSB-WC	Double-SideBand With-Carrier
DT	Discrete Time
FDM	Frequency-Division Multiplexing
FRF	Frequency Response Function
GCD	Greatest Common Divisor
IBPF	Ideal Band-Pass Filter
IBSF	Ideal Band-Stop Filter
IC	Initial Condition[s]
IHPF	Ideal High-Pass Filter
ILPF	Ideal Low-Pass Filter
IO	Input-Output
IR	Impulse Response
LCCDE	Linear Constant-Coefficient Differential Equation
LCM	Least Common Multiple
LHS	Left-Hand Side
LTI	Linear and Time-Invariant
NL	Nonlinear
OLHP	Open Left-Half Plane

OLTF	Open-Loop Transfer Function
ORHP	Open Right-Half Plane
PD	Proportional Derivative
PID	Proportional Integral Derivative
QAM	Quadrature Amplitude Modulation
Q.E.D.	Quod Erat Demonstrandum
RE	Reverse Engineering
RHS	Right-Hand Side
ROC	Region Of Convergence
SI	System Identification
SSB	Single-SideBand
TF	Transfer Function
TV	Time-Varying
ZIR	Zero-Input Response
ZSR	Zero-State Response

Chapter 1
Introduction

1.1 What Does "Signals and Systems" Mean?

Engineering systems are complex and perform sophisticated tasks involving signal processing, communications, and control. For example, consider a cellular phone. The sound wave received by the microphone has to be translated into an electric signal that is filtered, sampled, and quantized and the resulting digital signal transmitted to the nearest cellular tower using various modulation and multiplexing schemes. Similarly, the received digital signal has to be converted back to an analog signal, filtered, and amplified before being transformed into a sound wave by the speakers. Meanwhile, the software in the phone is running several algorithms (in the form of feedback loops) to control the transmitted power and maximize the battery lifetime, among other tasks.

Electrical and computer engineers need *formal techniques* to analyze, synthesize, and design the various components of engineering systems. The purpose of this book is to introduce, in a concise manner, these techniques, along with the associated mathematical transforms used in the analysis of linear time-invariant systems. Some simple applications in communications and control are also discussed.

1.1.1 What Is a Signal?

A signal is a function of an independent variable. When working in the "time domain" in this book, the independent variable will be *time*. The independent variable takes values in a domain set, and the function returns values in the codomain set. A signal could be a voltage or a current in an electrical circuit; it could be a velocity or a pressure in a mechanical system; it could be a binary signal that records a sequence of bits, as in digitized voice; it could be an epidemiological signal that counts the number of people infected with a certain virus; it could be a logical

© The Author(s), under exclusive license to Springer Nature Switzerland AG 2022
S. Lafortune, *A Guide to Signals and Systems in Continuous Time*,
https://doi.org/10.1007/978-3-030-93027-1_1

signal that records the packet types received by a router in a computer network; and so forth.

Signals are classified in terms of both their domains and codomains.

- Signal values (or codomain):

  - Continuum: continuous-amplitude signal (e.g., voltage)
  - Discrete set (finite or infinite): discrete-amplitude signals (e.g., population count), binary signals (0 or 1), logical signals

- Values of the independent variable (or domain), e.g., time:

  - Continuum: continuous-time signal (e.g., voltage, buffer contents)
  - Discrete set: discrete-time signal (e.g., monthly economic data)

All four combinations of the above are possible. Let us focus on signals over the "time" domain. Discrete-time signals arise in two ways: inherently discrete-time signals, e.g., economic data measured every month, or *sampled* continuous-time signals, e.g., music stored on your smart phone. Discrete-amplitude signals also arise in two ways: a signal can only take values in a discrete set or *quantized* continuous-amplitude signals, e.g., again, the music stored on your smart phone.

An *analog signal* is continuous-amplitude and continuous-time. A *digital signal* is discrete-amplitude and discrete-time. Analog-to-digital converters (ADCs) and digital-to-analog converters (DACs) transform signals from analog to digital and from digital to analog, respectively. A *discrete-event signal* is a continuous-time signal whose values are in a finite (discrete) set and whose transitions occur asynchronously over time, i.e., the transitions are event-driven as opposed to being time-driven.

Our focus in this book is on continuous-time (CT) signals over the (time) domain $\mathbb{R}$; these signals will be real-valued or complex-valued, i.e., the codomain will be $\mathbb{R}$ or $\mathbb{C}$.

1.1.2 What Is a System?

An engineering system is a "mechanism" (e.g., electro-mechanical) that operates on a signal (the input) and produces another signal (the output). We use the notation $x(t)$ for the input signal and the notation $y(t)$ for the output signal. There is a "cause–effect" relationship between the input and the output, as depicted in Fig. 1.1. In the context of this book, the word "system" will be used for the mathematical model of this possibly complex mechanism; thus, in the remainder of this book, "the model is the system." The cause–effect relationship from input to output will typically be modeled by a differential equation; e.g., $\frac{dy(t)}{dt} + ay(t) = bx(t)$ for some real constants a and b. It can also be modeled by a function that maps an input signal to an output signal; in simple cases, this function may involve only addition and multiplication operations, as in a modulator system that is described

Fig. 1.1 Black box view of a system, with input and output signals

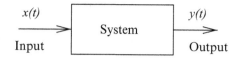

by $y(t) = x(t) \cos(\omega_0 t)$. The differential equation, or the explicit input-to-output function, is a mathematical abstraction of the actual mechanism of interest that is deemed accurate enough and allows the system to be analyzed formally. For example, a circuit with a resistor and a capacitor can be modeled by a first-order differential equation as written above; of course, many other systems may be modeled by the same differential equation. Hence, the analytical techniques studied in this book are generic and applicable to many classes of engineering systems.

Systems can be classified in a similar manner as was done for signals. Note however that there is no continuous-amplitude vs. discrete-amplitude distinction for systems.

- A *continuous-time system* (or CT system) is one that operates on continuous-time signals; it is often modeled by a *differential equation*. This book considers exclusively CT systems.
- A *discrete-time system* (or DT system) is one that operates on discrete-time signals; it is often modeled by a *difference equation*.
- A *discrete-event system* is one that operates on discrete-event signals; it is often modeled by a *finite-state machine*.

Let us assume that we have a system S that produces output $y(t)$ when it is excited by input $x(t)$. We define three problems that will be referred to throughout this book:

- *Problem IO:* Given $x(t)$ and a mathematical model of S, calculate $y(t)$.
  This is the standard "input–output" problem.
- *Problem RE:* Given $y(t)$ and a mathematical model of S, calculate $x(t)$.
  This is the "reverse engineering" problem. It may or may not have a [unique] solution.
- *Problem SI:* Given $x(t)$ and $y(t)$, derive a mathematical model of S that explains this behavior.
  This is the "system identification" problem. It may or may not have a [unique] solution.

For more complex systems composed of several interacting components, e.g., a demodulation system in an AM radio, it is customary to draw *block diagrams* that show in detail the interconnections between system components. System 1 and System 2 are connected in series when the output of System 1 is the input of System 2. In a parallel connection, the same input signal is applied to both System 1 and System 2 and the respective outputs are added. The feedback connection is common in control applications and it will be discussed later (see Sects. 6.1.2 and 6.6).

1.2   Preview of This Book

We now briefly overview some important notions that will be studied in this book to give the reader an insight about what is to come in our development. (This section may be skipped in a first reading.)

The Fourier and Laplace transforms, denoted by $\mathscr{F}$ and $\mathscr{L}$, respectively, are used to *map* analog signals and operations of interest between them from the time domain to the frequency domain (Fourier transform) and the s-domain (Laplace transform):

$$\text{Time Domain} \quad \overset{\mathscr{F}}{\longrightarrow} \quad \text{Frequency Domain}$$

$$\text{Time Domain} \quad \overset{\mathscr{L}}{\longrightarrow} \quad s\text{-Domain}$$

Recall *Euler's formula*:

$$e^{j\theta} = \cos(\theta) + j\sin(\theta)$$

from which we can write

$$\cos(\theta) = \frac{1}{2}e^{j\theta} + \frac{1}{2}e^{-j\theta} \quad \text{and} \quad \sin(\theta) = \frac{1}{2j}e^{j\theta} - \frac{1}{2j}e^{-j\theta}$$

Consider the complex plane with complex number

$$s = \sigma + j\omega = re^{j\theta} = r\cos(\theta) + jr\sin(\theta)$$

where $\sigma = \Re[s] = r\cos(\theta)$ and $\omega = \Im[s] = r\sin(\theta)$. (Notation: $\Re[z]$ denotes the real part and $\Im[z]$ denotes the imaginary part of complex number z.)

Let us consider the complex exponential function $e^{st} = e^{(\sigma+j\omega)t}$ and view s as a *frequency*. This is intuitively clear if $\sigma = 0$ since we get $e^{j\omega t}$, a periodic function of frequency ω, but let us take a "leap of faith" and think of s as a "generalized" frequency in e^{st}. In this book, we will do the following: (1) we will write signals as sums of complex exponential functions of the form $e^{s_i t}$ with their associated frequencies $s_i \in \mathbb{C}$ for a finite or infinite set of frequencies s_i. (2) We will look at how systems respond to each frequency s_i. (3) We will develop tools to switch between the time domain and the frequency domain (when $\sigma = 0$) or the s-domain (for general s), in order to obtain the time domain solution to the problem of interest.

Here are some examples of the "frequency representation" of signals:

1. Constant (often called DC) signal: $x(t) = 5$ for all $t \in \mathbb{R}$; frequency $s = 0$.
2. Sinusoidal signal: $x(t) = \cos(10t) = \frac{1}{2}e^{j10t} + \frac{1}{2}e^{-j10t}$; frequencies $s_1 = 10j$ and $s_2 = -10j$ or $\omega_1 = 10$ and $\omega_2 = -10$.
3. Exponential signal: $x(t) = e^{-3t}$; frequency $s = -3$.

4. Exponential multipied by sinusoid: $x(t) = e^{-3t} \cos(10t)$; frequencies $s_1 = -3 + 10j$ and $s_2 = -3 - 10j$.

The key reasons for being interested in the (generalized) frequency content of signals are the following:

1. We can (often) express a signal as a finite or infinite sum, or as an integral, of complex exponential functions of the form e^{st}.

   Periodic signals admit such representations; it is called the *Fourier Series* representation of the signal. In this case, $s = jk\omega_0$ for $k \in \mathbb{Z}$, i.e., s is purely imaginary and a multiple of the fundamental frequency ω_0 (cf. the $\cos(10t)$ example above).

   For aperiodic signals, we use the *Fourier transform* and/or the *Laplace transform*; both of these are representations of signals in terms of frequencies. In the case of the Fourier transform, $s = j\omega$, where $\omega \in \mathbb{R}$, while in the case of the Laplace transform, $s = \sigma + j\omega \in \mathbb{C}$.
2. Functions of the form e^{st} are *eigenfunctions* of continuous-time linear time-invariant (CT LTI) systems. This is a fundamental fact that we will prove in this book. (Roughly speaking, *linear* means that the principle of superposition holds and time-invariant means that the system parameters do not change values with time.)

The notion of an *eigenfunction* means the following:

$$\text{if } x(t) = e^{s_0 t}, \text{ then } y(t) = H(s_0)e^{s_0 t}$$

where $H(s_0)$ is a complex number that depends on s_0 (and on the system parameters). We can write $H(s_0)$ in complex exponential form to get

$$H(s_0) = |H(s_0)|e^{j \arg[H(s_0)]}$$

where $|H(s_0)|$ is the *magnitude* of the complex number $H(s_0)$ and $\arg[H(s_0)]$ is the *argument* or *phase* of the complex number $H(s_0)$. Thus, we have that

$$y(t) = |H(s_0)|e^{s_0 t + j \arg[H(s_0)]}$$

which means that the output is *also a complex exponential with the same frequency* s_0, but with a shift in phase by amount $\arg[H(s_0)]$ and a scaling in magnitude by amount $|H(s_0)|$.

The above result on eigenfunctions will be demonstrated after we establish the following result for the response of CT LTI systems:

$$y(t) = \int_{-\infty}^{\infty} h(t - \tau)x(\tau)d\tau$$

where $h(t)$ is a function that characterizes the system; it is called the *impulse response* of the system. This integral is called the *convolution integral*. Moreover, $\mathcal{L}[h(t)] = H(s)$, where $H(s)$ (a function of s) is called the *Transfer Function* of the system. For "stable" systems, we will also define $\mathcal{F}[h(t)] = H(j\omega)$, the *Frequency Response Function* of the system.

By working in the frequency or s-domain, we will be able to map the convolution integral to a *multiplication* operation between the [Fourier or Laplace] transforms of interest. This will greatly simplify the solution to Problem IO, and it will also allow, in most cases, to solve Problems RE and SI as well.

This concludes this brief "preview of the book" section.

1.3 Properties of Signals

We present some background material on signals, which will be used in the remainder of the book. The reader is encouraged to consult Appendix A, where common engineering (analog) signals are reviewed.

1.3.1 Symmetry and Periodicity

Let $f(t)$ be a CT function (or signal in our terminology).
Even symmetry: $f(t)$ is said to be even if it is symmetric with respect to the vertical axis, i.e., $f(t) = f(-t)$ for all $t \in \mathbb{R}$.
Odd symmetry: $f(t)$ is said to be odd if it is symmetric with respect to the origin, i.e., $f(t) = -f(-t)$ for all $t \in \mathbb{R}$.
Periodicity: $f(t)$ is periodic with period T if

$$f(t + kT) = f(t) \quad \text{for all } t \in \mathbb{R} \quad \text{and for all } k \in \mathbb{Z}$$

The smallest such T is called the *fundamental period* and is usually denoted by T_0.

Observe that by definition, a periodic function should start at $t = -\infty$ and end at $t = \infty$. If T is the period (in seconds), then $f = \frac{1}{T}$ is the *frequency* (in Hz or cycles per second) and $\omega = 2\pi f = \frac{2\pi}{T}$ is the *radian frequency* (in radians per second). The *average value* of a periodic function with period T is

$$f_{avg} = \frac{1}{T} \int_{t_0}^{t_0+T} f(t)dt$$

for any $t_0 \in \mathbb{R}$. The average value is often called the "DC component" of the signal.

Two important classes of periodic functions used extensively in this book are (1) $e^{jk\omega_0 t}$ and (2) $\cos(k\omega_0 t)$, where in both cases $k \in \mathbb{Z}$ and $\omega_0 > 0$ by convention. Further details about these signals are given in Appendix A. These two families of signals are said to be *harmonically related*; the frequency ω_0 is called the *fundamental frequency*, while the frequency $k\omega_0$ for $k > 1$ is called the *kth harmonic*.

As a final observation, we note that the sum of two periodic functions $x_1(t)$ and $x_2(t)$, with fundamental periods T_1 and T_2, respectively, will be periodic if $\frac{T_1}{T_2}$ is a rational number. In this case, the fundamental period of the resulting signal will be the least common multiple (LCM) of (real numbers) T_1 and T_2; equivalently, the greatest common divisor (GCD) of $\frac{1}{T_1}$ and $\frac{1}{T_2}$ will be the fundamental frequency (in Hz).

1.3.2 Notions of Energy and Power Signals

It is common to classify signals according to the two notions of *energy* and *power*. To motivate these definitions in terms of a concrete example, let $f(t)$ be the voltage across a 1 ohm resistor; then, $|f(t)|^2$ is the instantaneous power delivered by $f(t)$. Accordingly, the definitions of energy and power signals are as follows, where in general $f(t)$ may be complex (in which case $|f(t)|$ means the magnitude of this complex number):

- Energy of $f(t)$ in $[T_1, T_2]$:

$$E_{[T_1,T_2]} = \int_{T_1}^{T_2} |f(t)|^2 dt$$

- Total energy in $f(t)$:

$$E_{total} = \lim_{T \to \infty} E_{[-\frac{T}{2}, \frac{T}{2}]}$$

Note that this integral may go to ∞. If E_{total} of $f(t)$ is finite, then $f(t)$ is called an *energy signal*.

Bounded finite-duration signals are energy signals. Some infinite-duration signals are energy signals. Take $f(t) = e^{-a|t|}$ with $a > 0$. Then, verify that $E_{total} = \frac{1}{a}$.

- Average power of $f(t)$ in $[T_1, T_2]$:

$$P_{[T_1,T_2]}^{avg} = \frac{1}{T_2 - T_1} E_{[T_1,T_2]} = \frac{1}{T_2 - T_1} \int_{T_1}^{T_2} |f(t)|^2 dt$$

- Average power of $f(t)$:

$$P^{avg} = \lim_{T \to \infty} \frac{1}{T} \int_{-\frac{T}{2}}^{\frac{T}{2}} |f(t)|^2 dt$$

If a signal has infinite energy, but P^{avg} is finite, then it is called a *power signal*. An energy signal will yield $P^{avg} = 0$.

- Periodic signals are power signals if they are bounded. In this case,

$$P^{avg} = \frac{1}{T_0} \int_{-\frac{T_0}{2}}^{\frac{T_0}{2}} |f(t)|^2 dt$$

$f(t) = A \sin(\omega_0 t)$ is a power signal with $P_{avg} = \frac{A^2}{2}$.

- Some signals are neither energy nor power signals; one such example is the ramp signal $f(t) = r(t) = tu(t)$.

1.4   Signal Transformations

This section may be skipped until Sect. 2.2, when the transformations below will be employed.

We often need to manipulate signals in terms of their amplitude or in terms of the independent variable. For instance, in the expression of the *convolution integral*, the *impulse response* appears in the form $h(t - \tau)$ where in that context τ is the *independent* variable and t is some fixed "shift":

$$y(t) = \int_{-\infty}^{\infty} h(t - \tau)x(\tau)d\tau$$

That is, to calculate the output at time t_1, we must do

$$y(t_1) = \int_{-\infty}^{\infty} h(t_1 - \tau)x(\tau)d\tau = \int_{-\infty}^{\infty} h(t_1 - t)x(t)dt$$

In the above, $h(-t + t_1) = h(-(t - t_1))$ is a timed-transformed version of $h(t)$ involving time reversal (the minus sign) and time shifting (the $-t_1$ term).

Time Reversal: $x_{new}(t) = x_{old}(-t)$.
The graph of $x_{new}(t)$ vs. t is obtained by taking the mirror image of the graph of $x_{old}(t)$ vs. t with respect to the vertical axis ("reflection").

Time Shifting: $x_{new}(t) = x_{old}(t - t_0)$, where t_0 is a constant.
The graph of $x_{new}(t)$ vs. t is obtained by *shifting* the graph of $x_{old}(t)$ vs. t to the right (delay) by amount t_0 if t_0 is positive or to the left (prediction) by amount $|t_0|$ if t_0 is negative.

Time Scaling: $x_{new}(t) = x_{old}(at)$, where a is a positive constant.
The graph of $x_{new}(t)$ vs. t is obtained by *expanding* the graph of $x_{old}(t)$ vs. t by a factor of $\frac{1}{a}$ if $0 < a < 1$ or by *compressing* the graph of $x_{old}(t)$ vs. t by a factor of a if $a > 1$.

1.4.1   Combinations of Time Transformations

The following procedure is recommended when there is a combination of scaling and time shift:

1. Write

$$x_{new}(t) = x_{old}(at + b) = x_{old}(a(t + \frac{b}{a}))$$

2. Scale the plot of $x_{old}(t)$ vs. t by amount $|a|$; the scaling will be an expansion if $0 < |a| < 1$ and a compression if $|a| > 1$.
3. Reflect the resulting plot if $a < 0$.
4. Shift the resulting plot by amount $|\frac{b}{a}|$, where the shift is to the right if $\frac{b}{a} < 0$ and to the left if $\frac{b}{a} > 0$.

Note that the order of the operations is important in the above procedure since shift does not commute with the other operations.

   Let us return to the convolution integral and consider obtaining the plot of $h(-t + t_1)$ vs. t from the plot of $h(t)$ vs. t. $h(-t + t_1) = h(-(t - t_1))$; hence, the original plot is first reflected and then shifted by amount $|t_1|$, with a right shift if $t_1 > 0$ and a left shift if $t_1 < 0$.

Chapter 2
Systems

2.1 Properties of Systems

Throughout this section, we will use the following illustrative examples of systems:

1. Amplitude Modulation [AM]:
   Input: $x(t)$
   Output: $y(t) = x(t)\cos(\omega_0 t)$
2. Voltage Limiter [VL]:
   Input: $x(t)$
   Output:

$$y(t) = \begin{cases} x(t) & \text{if } |x(t)| \leq B \\ \text{sgn}(x(t)) \cdot B & \text{otherwise} \end{cases}$$

3. Average [Av]:
   Input: $x(t)$
   Output: $y(t) = \frac{1}{2}[x(t-1) + x(t+1)]$
4. Series RL Circuit with Voltage Source [RL]:
   Input: $x(t)$ is the voltage applied to the circuit with R and L in series.
   Output: $y(t)$ is the resulting current in the circuit.

2.1.1 Memorylessness and Causality

Systems with/without Memory A system is *static* or *zero-memory* or *memoryless* if, for all $t_1 \in \mathbb{R}$, the value of the output at time t_1, $y(t_1)$, depends only on the value of the input at the same time, $x(t_1)$, and not on any other values of the input function;

© The Author(s), under exclusive license to Springer Nature Switzerland AG 2022
S. Lafortune, *A Guide to Signals and Systems in Continuous Time*,
https://doi.org/10.1007/978-3-030-93027-1_2

in other words, $y(t_1)$ does not depend on $x(t)$ for $t \neq t_1$. A system that is not static is called *dynamic*.

AM and VL are static, while Av and RL are dynamic.

Causality A system is *causal* or *non-anticipatory* if, for all $t_1 \in \mathbb{R}$, the value of the output at time t_1, $y(t_1)$, depends only on past and current values of the input function, i.e., $x(t)$ for $t \leq t_1$, and not on any *future* values of the input function, i.e., $x(t)$ for $t > t_1$.

All physical ("real-life") systems are causal. Offline processing of signals can be non-causal. All static systems are causal by definition.

Av is not causal, while AM, VL, and RL are causal.

Notion of State of Dynamic System In general, for dynamic systems, the calculation of the output of the system requires not only the input function but also potential "initial conditions" (IC) that are present at the moment when the input is first applied. It is customary to refer to the set of required initial conditions at time t_0 as the *state* of the system at time t_0, denoted by $s(t_0)$. An abstract way to define $s(t_0)$ is to say that it is the "minimum amount of information that summarizes all past behavior" at time t_0. That is, if we know

$$s(t_0) \ \text{ and } \ x(t) \text{ for } t \geq t_0$$

then we can calculate

$$y(t) \text{ for } t \geq t_0$$

The conceptual description of a system with explicit representation of the initial state is shown in Fig. 2.1.

Causal dynamic systems are often described by differential equations. In this case, the state is the set of initial conditions necessary to solve the differential equation. The state is a scalar for first-order systems; otherwise, it is a vector. In the case of RL, the state $s(t_0)$ is the current in the inductor, which is the initial condition for the differential equation. More generally, in an RLC circuit, $s(t_0)$ would be the values of the currents in the inductors and of the voltages across the capacitors at time t_0.

Fig. 2.1 Black box view of a system, with input and output signals and with initial state (or initial conditions)

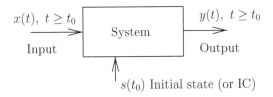

2.1.2   Zero-State Response and Zero-Input Response

Consider a dynamic system, where an input is applied starting at time t_0; let $s(t_0)$ be the state of the system at time t_0. In general, $y(t)$ for $t \geq t_0$ depends on both the initial state $s(t_0)$ and the input function $x(\tau)$, $\tau \geq t_0$. We write

$$y(t) = f(s(t_0); \; x(\tau), \; \tau \geq t_0)$$

We define two special cases of this output.

- The *Zero-Input Response*, denoted by $ZIR(t)$, is the response (output) due to $s(t_0)$ when $x(\tau) = 0$ for all $\tau \geq t_0$:

$$ZIR(t) = f(s(t_0); \; 0)$$

- The *Zero-State Response*, denoted by $ZSR(t)$, is the response (output) due to the input function $x(\tau)$, $\tau \geq t_0$, when $s(t_0) = 0$:

$$ZSR(t) = f(0; \; x(\tau), \; \tau \geq t_0)$$

In general, it is not true that the total output is the sum of the ZSR and the ZIR; for example, consider the system:

$$y(t) = \left[s(t_0) + \int_{t_0}^{t} x(\tau) d\tau \right]^2$$

A sufficient condition for being able to decompose the total output into the sum of the ZIR and the ZSR is presented in Sect. 2.1.5.

At this point, it is useful to present the general solution of a particular first-order differential equation, as we will return to this type of equation frequently and it will help to guide the discussion.

First-Order Differential Equation   Consider the first-order differential equation

$$\frac{dy(t)}{dt} + ay(t) = bx(t) \text{ for } t \geq t_0 \tag{2.1}$$

with initial condition $y(t_0)$. We claim that the solution of this differential equation, for any $x(t)$, is given by the expression:

$$y(t) = e^{-a(t-t_0)} y(t_0) + \int_{t_0}^{t} e^{-a(t-\tau)} bx(\tau) d\tau , \quad t \geq t_0 \tag{2.2}$$

In particular, if $t_0 = 0$, we get that

$$y(t) = e^{-at}y(0) + \int_0^t e^{-a(t-\tau)}bx(\tau)d\tau , \quad t \geq 0 \tag{2.3}$$

Note that the first term depends only on the initial condition $y(t_0)$ and that the second term depends only on the input signal $x(t)$ (sometimes called the "forcing function"). Thus, the first term is the ZIR and the second term is the ZSR, and in this case it is true that the total output is the sum of the ZIR and the ZSR. Later on, we will refer to the function e^{-at} inside the integral of the ZSR as the *mode* of this system.

To verify that the expression in Eq. (2.2) is correct, we first observe that it satisfies the initial condition at time t_0. Second, we verify that it satisfies the differential equation (2.1) itself; this step requires applying the Leibniz[1] integral rule when doing $\frac{d}{dt}$ of the integral term in Eq. (2.2). The detailed steps are left to the reader.

RL is described by a first-order differential equation of the form in Eq. (2.1). Let the initial condition, which is the current in the inductor, be $y(t_0)$ at the initial time t_0. From basic circuit theory, for $t \geq t_0$,

$$x(t) = Ry(t) + L\frac{dy(t)}{dt}$$

which can be rewritten as

$$\frac{dy(t)}{dt} + \frac{R}{L}y(t) = \frac{1}{L}x(t)$$

2.1.3   Stability

We introduce two notions of stability for systems.

BIBO Stability Bounded-Input Bounded-Output (BIBO) stability means that if the input does not "blow up," then the output will not "blow up." Formally, we set $s(t_0) = 0$. Then, the system is *BIBO stable* if

$$|x(t)| \leq M_x < \infty \text{ for all } t \geq t_0 \Rightarrow |y(t)| \leq M_y \text{ for all } t \geq t_0, \text{ where } M_y < \infty$$

(Note that M_x and M_y are *constants*.)

AM, VL, and Av are all BIBO stable by inspection. For RL, we note that $a = \frac{R}{L}$ in Eq. (2.3); we will see an easy test later on that proves BIBO stability.

---

[1] Gottfried Wilhelm Leibniz was a German mathematician and scientist, 1646–1716.

Asymptotic Stability A dynamic system is said to be *asymptotically stable* if

$$\lim_{t\to\infty} ZIR(t) = 0 \text{ for all } s(t_0)$$

RL is asymptotically stable, since it is a passive circuit.

2.1.4 Time-Invariance and Linearity

Time-Invariance Informally, a system is *time-invariant* (TI) if the values of its parameters do not change with time. A system described by a differential equation is time-invariant if all the coefficients of the terms involving the functions $y(t)$ and $x(t)$ do not depend on t.

To define time-invariance formally, we start by writing

$$y_1(t) = f(s_1(t_0); \ x_1(\tau), \ \tau \geq t_0), \ t \geq t_0$$
$$y_2(t) = f(s_2(t_0 + t_d); \ x_2(\tau), \ \tau \geq t_0 + t_d), \ t \geq t_0 + t_d$$

Then, the system is time-invariant if

$$s_2(t_0 + t_d) = s_1(t_0) \quad \text{and} \quad x_2(t) = x_1(t - t_d) \text{ for } t \geq t_0 + t_d$$

$$\text{imply}$$

$$y_2(t) \quad = \quad y_1(t - t_d) \text{ for } t \geq t_0 + t_d$$

for all time shifts $t_d \in \mathbb{R}$, for all initial states $s_1(t_0)$, and for all inputs $x_1(t)$. A system that is not time-invariant is called *time-varying* (TV).

AM is TV, while VL, Av, and RL are TI.

Linearity To formally define linearity of systems in the presence of initial conditions, we proceed as follows. Let us write

$$y_1(t) = f(s_1(t_0); \ x_1(\tau), \ \tau \geq t_0), \ t \geq t_0$$
$$y_2(t) = f(s_2(t_0); \ x_2(\tau), \ \tau \geq t_0), \ t \geq t_0$$

Then, the system is *linear* if

$$f(\alpha_1 s_1(t_0) + \alpha_2 s_2(t_0); \alpha_1 x_1(\tau) + \alpha_2 x_2(\tau), \tau \geq t_0) = \alpha_1 y_1(t) + \alpha_2 y_2(t)$$

for all $t \geq t_0$, for all real or complex constants α_1 and α_2, for all initial states $s_1(t_0)$ and $s_2(t_0)$, and for all inputs $x_1(t)$ and $x_2(t)$. The sum property is known as *superposition* or *additivity*, while the scaling property is known as *homogeneity*.

AM, Av, and RL are linear, while VL is nonlinear.

A system described by a differential equation of the form

$$\sum_{i=0}^{n} a_i \frac{d^i y(t)}{dt^i} = \sum_{i=0}^{m} b_i \frac{d^i x(t)}{dt^i}$$

is linear. It will also be time-invariant if the a's and b's are constant. In that case, we say that we have a linear constant-coefficient differential equation (LCCDE). In this book, we focus almost exclusively on LTI systems that are modeled by LCCDEs.

Simpler Definitions of Linearity and Time-Invariance For the sake of generality, linearity and time-invariance were defined above accounting for the initial state of the system. In particular, linearity was defined *jointly* with respect to the initial state (or IC) and the input. It is also common to define linearity and time-invariance *assuming that the initial state (or IC) is equal to 0*. Our motivation for the more general definition is seen in the next section.

2.1.5 Complete Response of LTI Systems

Consider a linear system. Take

$$y_1(t) = f(s_1(t_0); \, 0), \, t \geq t_0$$

$$y_2(t) = f(0; \, x_2(\tau), \, \tau \geq t_0), \, t \geq t_0$$

i.e., $x_1(\tau) = 0$ for $\tau \geq t_0$ and $s_2(t_0) = 0$. Then, taking $\alpha_1 = \alpha_2 = 1$ and applying linearity, we get

$$f(s_1(t_0); \, x_2(\tau), \, \tau \geq t_0) = y_1(t) + y_2(t) = f(s_1(t_0); \, 0) + f(0; \, x_2(\tau), \, \tau \geq t_0)$$

This means that

$$\text{Complete Response of Linear System} = \text{ZIR} + \text{ZSR}$$

For this reason, we will focus on the computation of the ZSR in the remainder of this chapter, as well as in Chaps. 3 and 4. We will only return to the ZIR when we use the Laplace transform to solve LCCDEs in Chap. 6.

2.2 LTI Systems: Time-Domain Analysis

In the remainder of this book, we will focus primarily on (CT) LTI systems. It is important to understand that while few physical systems are exactly LTI, it is often possible to make LTI approximations of physical systems that are adequate for the ensuing work; such approximations are obtained by *linearization*.

In this section, we analyze LTI systems in the *time domain*. Namely, we ask the question:

Given an LTI system and given the input $x(t)$ that is applied to the system after t_0, how can we calculate the output $y(t)$ for $t \geq t_0$?

Observe that unless otherwise stated, *we assume that $s(t_0) = 0$ hereafter*, i.e., we focus our attention on the ZSR. By linearity, we know that the complete response will be the sum of the ZSR with the ZIR. By time-invariance, we can set $t_0 = 0$, unless $t_0 = -\infty$.

At this point, we need to define a special signal: the unit impulse signal.

2.2.1 Unit Impulse Signal

Definition of Unit Impulse Signal

We now define a special "function" that is called the *unit impulse* [signal] or *Delta function* or *Dirac*[2] *function*. It is denoted by $\delta(t)$. Mathematically speaking, $\delta(t)$ is not a function, but it is a generalized function or singularity function or distribution. This "function" is very useful in engineering, and as long as we are aware of what we are doing, it is acceptable to have a certain measure of "mathematical informality" in the subsequent discussion.

Engineering Interpretation We can "define" $\delta(t)$ as a "function" that satisfies the following two properties:

1. $\delta(t) = 0$ for all t except for $t = 0$.
2. The total area under $\delta(t)$ is 1, i.e.,

$$\int_{-\infty}^{\infty} \delta(t)dt = 1$$

Pictorially, we draw $\delta(t)$ as a spike at the origin and write "1" next to it to indicate the total area under this spike. Thus, the plot of $3\delta(t - 5)$ is a spike of area (or weight) 3 at $t = 5$.

---

[2] Paul Adrien Maurice Dirac was an English physicist, 1902–1984.

Definition by Limit A mathematical justification of the preceding definition is now given.

Let $p_1(t) = \text{rect}(t)$.
Let $p_2(t) = 2 \cdot \text{rect}(t/0.5) = 2 \cdot \text{rect}(2t)$.
Let $p_n(t) = n \cdot \text{rect}(nt)$.

Clearly, the total area under each function p_n is 1. Note that each $p_n(t)$ is piecewise continuous and even. Then, we can view $\delta(t)$ as

$$\delta(t) = \lim_{n \to \infty} p_n(t)$$

Observe that the limit satisfies the two conditions in the "engineering interpretation" of $\delta(t)$.

Other functions than rectangular pulses could be used in the preceding argument. For example, we could take the *Gaussian pulses*:

$$p'_n(t) = \frac{n}{\sqrt{2\pi}} \exp\left(\frac{-t^2 n^2}{2}\right)$$

Properties of Unit Impulse Signal

The properties below can be proved using the "definition by limit" of the unit impulse.

- *Evenness:* $\delta(t) = \delta(-t)$.
- *Sifting property:*

$$\int_{t_1}^{t_2} f(t)\delta(t - t_0)dt = \begin{cases} f(t_0) & \text{if } t_1 < t_0 < t_2 \\ 0 & \text{otherwise} \end{cases}$$

  as long as $f(t)$ is continuous at $t = t_0$.
- *Identity element of convolution:*

$$f(t) = \int_{-\infty}^{\infty} f(\tau)\delta(t - \tau)d\tau$$

  for all t where $f(t)$ is continuous.

  This result follows from the sifting property. Later on, we will see that this integral is called a *convolution integral*, hence the name of this property. Observe that this result can also be used to define the unit impulse.

- *Sampling property:*

$$f(t)\delta(t - t_0) = f(t_0)\delta(t - t_0)$$

  as long as $f(t)$ is continuous at $t = t_0$.

- *Scaling property:*

$$\delta(at + b) = \frac{1}{|a|}\delta\left(t + \frac{b}{a}\right)$$

Observe that each side of this equation is a spike at $t = -\frac{b}{a}$ of area $\frac{1}{|a|}$.
- *Relation with unit step:* From the engineering definition of the unit impulse, we have that

$$\int_{-\infty}^{t} \delta(\tau)d\tau = \begin{cases} 1 & \text{if } t > 0 \\ 0 & \text{if } t < 0 \end{cases} = u(t)$$

Thus, we can write

$$\frac{du(t)}{dt} \text{ "} = \text{" } \delta(t)$$

The above derivative relationship between the unit step and the unit impulse is only meaningful as a limiting argument using the previously defined sequence of functions $p_n(t)$, as we now explain.

1. Consider the sequence of functions $u_n(t)$ where

$$\frac{du_n(t)}{dt} = p_n(t)$$

Each $u_n(t)$ is a ramp-like function of slope n from $-\frac{1}{2n}$ to $\frac{1}{2n}$ and it is equal to 0 before $-\frac{1}{2n}$ and equal to 1 after $\frac{1}{2n}$.
2. Observe that

$$\lim_{n \to \infty} u_n(t) = u(t)$$

3. Since

$$\lim_{n \to \infty} p_n(t) = \delta(t)$$

we conclude that

$$\frac{du(t)}{dt} = \delta(t)$$

Hence, the above claim is substantiated by the limiting argument.

2.2.2 Impulse Response

We define the *Impulse Response* (IR) of a system, denoted by $h(t)$, to be the output
of the system due to input $x(t) = \delta(t)$, when $s(0) = 0$, i.e.,

$$h(t) = f(0; \delta(t)) = \text{ZSR due to } \delta(t)$$

We present two examples.

1. Let us return to the first-order LCCDE (2.1), whose solution is given in Eq. (2.3).
   We use Eq. (2.3) to derive the impulse response of a first-order LCCDE:

$$h(t) = \begin{cases} 0 & \text{if } t < 0 \\ \int_{0^-}^{t} e^{-a(t-\tau)} b\delta(\tau)d\tau & \text{if } t > 0 \end{cases}$$

$$= e^{-at} bu(t)$$

   using the sifting property. We need to write 0^- as the lower limit of the integral
   to make sure that the unit impulse falls in the range of the integral. Also, the $u(t)$
   in the expression of $h(t)$ indicates that $h(t) = 0$ whenever $t < 0$. (If we need a
   value for $h(0)$, we set $h(0) = b$.)
2. The impulse response of Av is $h(t) = \frac{1}{2}[\delta(t-1) + \delta(t+1)]$.

2.2.3 Convolution Integral Theorem

Convolution Integral Theorem: Given the impulse response $h(t)$ of an LTI
system, we can calculate the ZSR due to any input function $x(t)$ by doing the
following *convolution integral*:

$$ZSR(t) = \int_{-\infty}^{\infty} h(t-\tau)x(\tau)d\tau \triangleq h(\cdot) * x(\cdot)|_t \equiv h(t) * x(t)$$

where $\triangleq$ means "equal to by definition." The notation $h(\cdot) * x(\cdot)|_t$ would be
preferable to the notation $h(t) * x(t)$, since convolution is an operation on the entire
time functions $x(\cdot)$ and $h(\cdot)$, where the parameter t in the convolution integral is the
time at which the output is calculated. Thus, it is not a static operation involving only
the values of these functions at time t. Nevertheless, we will often write $h(t) * x(t)$
as it is common practice in most textbooks. In the convolution integral, the variable
of integration is τ; thus, $h(t-\tau) = h(-(\tau-t))$ means that the function $h(\cdot)$ is first
reflected and then shifted to the right by amount t (for positive values of t). Recall
our discussion of signal transformations in Sect. 1.4.

The proof of the Convolution Integral Theorem[3] is given in Appendix B.1.1.[4] We encourage the reader to read the proof and verify that both linearity and time-invariance need to be invoked. The Convolution Integral Theorem is a fundamental result for LTI systems, as it shows that knowledge of the IR suffices to calculate the ZSR for any input to the system.

Next, we present some examples of convolution.

1. *First-Order LCCDE:* The impulse response of the LCCDE in Eq. (2.1) is $h(t) = e^{-at}bu(t)$, as was shown above. If we apply input $x(t)$ starting at $t = 0$, then $x(t) = x(t)u(t)$. Substituting in the convolution integral, we get, for $t \geq 0$,

$$ZSR(t) = \int_{-\infty}^{\infty} h(t - \tau)x(\tau)d\tau$$

$$= \int_{-\infty}^{\infty} e^{-a(t-\tau)}bu(t - \tau)x(\tau)u(\tau)d\tau$$

$$= \int_{0}^{t} e^{-a(t-\tau)}bx(\tau)d\tau$$

where the range of the integral has been adjusted to account for when the unit step functions in the integrand evaluate to 0. This expression agrees with our earlier result in Eq. (2.3).

For example, if $x(t) = u(t)$, we get after integration that

$$ZSR(t) = \frac{b}{a}[1 - e^{-at}]u(t)$$

where the $u(t)$ term indicates that $ZSR(t) = 0$ if $t < 0$.

2. *Integrator:* Consider an integrator, i.e., a system where

$$y(t) = \int_{-\infty}^{t} x(\tau)d\tau$$

The impulse response of this system is

$$h(t) = \int_{-\infty}^{t} \delta(\tau)d\tau = u(t)$$

Indeed,

$$y(t) = \int_{-\infty}^{\infty} u(t - \tau)x(\tau)d\tau = \int_{-\infty}^{t} x(\tau)d\tau$$

---

[3] The terminology "Convolution Theorem" usually refers to the convolution *property* of Fourier and Laplace transforms, which states that convolution in the time domain is multiplication in the frequency and s-domains, respectively, as we will see in Sects. 4.2.1 and 6.1.2.

[4] Most of the proofs in this book have been grouped in Appendix B.

3. *Response to Unit Step:* If we apply $x(t) = u(t)$ to an LTI system, then the resulting output (ZSR) is called the *step response* and denoted by $y_{step}(t)$. We have that

$$y_{step}(t) = \int_{-\infty}^{\infty} h(\tau)u(t-\tau)d\tau = \int_{-\infty}^{t} h(\tau)d\tau$$

Thus, if we *differentiate* the step response, we obtain the impulse response:

$$\frac{dy_{step}(t)}{dt} = h(t)$$

This is a useful way of experimentally obtaining the impulse response.

The convolution integral provides a complete solution of Problem IO (for the ZSR). However, there is no "invertibility" mechanism for the convolution integral, so we do not have a general solution procedure for Problem RE and for Problem SI at this time. This issue will be resolved in later chapters when we work in the frequency and s-domains.

2.2.4 Properties of Convolution

Commutativity:

$$h(\cdot) * x(\cdot)|_t = \int_{-\infty}^{\infty} h(t-\tau)x(\tau)d\tau = \int_{-\infty}^{\infty} x(t-\sigma)h(\sigma)d\sigma \quad [\sigma = t-\tau]$$

$$= x(\cdot) * h(\cdot)|_t$$

Associativity:

$$[x(\cdot) * h_1(\cdot)] * h_2(\cdot)|_t = x(\cdot) * [h_1(\cdot) * h_2(\cdot)]|_t$$

This implies that the impulse response of the series connection of two systems is the convolution of the individual impulse responses. Associativity is proved as follows:

$$y_1(t) = \int_{-\infty}^{\infty} x_1(\tau)h_1(t-\tau)d\tau \implies y_1(t-\sigma) = \int_{-\infty}^{\infty} x_1(\tau)h_1(t-\sigma-\tau)d\tau$$

$$y(t) = \int_{-\infty}^{\infty} h_2(\sigma)y_1(t-\sigma)d\sigma = \int_{-\infty}^{\infty} h_2(\sigma)\left[\int_{-\infty}^{\infty} x_1(\tau)h_1(t-\sigma-\tau)d\tau\right]d\sigma$$

$$= \int_{-\infty}^{\infty} x_1(\tau)\left[\int_{-\infty}^{\infty} h_2(\sigma)h_1(t-\sigma-\tau)d\sigma\right]d\tau = \int_{-\infty}^{\infty} x_1(\tau)h(t-\tau)d\tau$$

where $h(t) \overset{\Delta}{=} h_1(\cdot) * h_2(\cdot)|_t$.

Distributivity:

$$x(\cdot) * [h_1(\cdot) + h_2(\cdot)]|_t = [x(\cdot) * h_1(\cdot)] + [x(\cdot) * h_2(\cdot)]|_t$$

This follows immediately from the properties of integrals. Thus the impulse response of the parallel connection of two systems is the sum of their individual impulse responses.

Identity Element: From the sifting property of $\delta(t)$,

$$x(\cdot) * \delta(\cdot)|_t = \int_{-\infty}^{\infty} x(\tau)\delta(t - \tau)d\tau = x(t)$$

In words, the unit impulse is the identity element of convolution.

From the properties of the unit impulse and the definition of convolution, we have the following useful result:

$$x(t - t_a) * \delta(t - t_b) = \int_{-\infty}^{\infty} x(\tau - t_a)\delta(t - \tau - t_b)d\tau = x(t - t_a - t_b) \qquad (2.4)$$

Using time-invariance, a general form of this result is

$$f_1(t) * f_2(t) = g(t) \quad \Rightarrow \quad f_1(t - t_1) * f_2(t - t_2) = g(t - t_1 - t_2)$$

Table 2.1 presents the answer to a few common convolutions. The preceding properties can be leveraged to extend this knowledge base.

2.2.5 Graphical Convolution

The convolution integral is sometimes a source of confusion due to the use of both t and τ. Let us calculate the convolution of the two functions $x(\cdot)$ and $h(\cdot)$ at time $t = t_1$:

$$y(t_1) = \int_{-\infty}^{\infty} h(t_1 - \tau)x(\tau)d\tau$$

Table 2.1 Results of a few common convolution integrals ($a \in \mathbb{R}$)

$f(t)$	$g(t)$	$f(t) * g(t)$
$e^{-at}u(t)$	$e^{-bt}u(t)$	$\frac{1}{b-a}[e^{-at} - e^{-bt}]u(t)$ if $a \neq b$
$e^{-at}u(t)$	$e^{-at}u(t)$	$te^{-at}u(t)$
$te^{-at}u(t)$	$e^{-at}u(t)$	$\frac{1}{2}t^2e^{-at}u(t)$

We use t_1 here to highlight the fact that τ is the actual variable of integration (of course, we could keep using t!). When $x(\cdot)$ and $h(\cdot)$ have analytical expressions that contain several terms (e.g., because of discontinuities), the range of the integral might need to be adjusted for different values of t_1. As a simple example, let $x(t) = h(t) = u(t) - u(t-1)$. Then, we have to evaluate

$$y(t_1) = \int_{-\infty}^{\infty} [u(t_1 - \tau) - u(t_1 - \tau - 1)][u(\tau) - u(\tau - 1)]d\tau \qquad (2.5)$$

which means that the bounds of the integral will change for different values of t_1, depending on where the combination of the four unit step functions in the integrand evaluate to 0 or 1. For instance, if $t_1 = 1$, then Eq. (2.5) reduces to

$$y(1) = \int_0^1 1 d\tau = 1$$

To "visualize" this, and to treat similar examples, it is often convenient to plot $x(\tau)$ and $h(t_1 - \tau)$, for a fixed t_1; this allows determining what values the product $x(\tau)h(t_1 - \tau)$ takes (for a fixed t_1 and for all τ) more easily (hopefully!) than by examining the analytical expressions of the functions $x(\cdot)$ and $h(\cdot)$. Then, we can calculate the area under the product of these two functions (of τ) in order to obtain the value of $x(\cdot) * h(\cdot)|_{t_1}$.

We suggest the following procedure to perform the convolution integral "graphically":

1. Plot $x(\tau)$ vs. τ.
2. Flip $h(\tau)$ w.r.t. the vertical axis to get $h(-\tau)$ vs. τ.
3. To get $h(t_1 - \tau) = h(-(\tau - t_1))$, shift $h(-\tau)$ by amount t_1 in the following manner: *a positive t_1 means shifting $h(-\tau)$ to the right and a negative t_1 means shifting $h(-\tau)$ to the left.*
4. By sliding $h(t_1 - \tau)$ over $x(\tau)$, identify all the relevant intervals of t_1 where the analytical expression of the non-zero part of the overlap of these functions changes.
5. For each interval of t_1 identified above, calculate the area under the product $x(\tau)h(t_1 - \tau)$, as a function of t_1, i.e., write the exact expression of the convolution integral in this interval and calculate the integral.

In our simple example where $x(t) = h(t) = u(t) - u(t-1)$, there are four relevant intervals for t_1:

1. $t_1 < 0$, for which $x(\tau)h(t_1 - \tau) = 0$ for all τ.
2. $0 < t_1 < 1$, for which $x(\tau)h(t_1 - \tau) = 1$ for $0 < \tau < t_1$.
3. $1 < t_1 < 2$, for which $x(\tau)h(t_1 - \tau) = 1$ for $t_1 - 1 < \tau < 1$.
4. $t_1 > 2$, for which $x(\tau)h(t_1 - \tau) = 0$ for all τ.

The complete answer is (where we return to the usual t notation by replacing t_1 by t!)

$$y(t) = x(\cdot) * h(\cdot)|_t = \begin{cases} 0 & \text{for } t \leq 0 \\ t & \text{for } 0 \leq t \leq 1 \\ 2 - t & \text{for } 1 \leq t \leq 2 \\ 0 & \text{for } t \geq 2 \end{cases}$$

2.2.6 System Properties and the Impulse Response

Since an LTI system is completely described by its impulse response (up to the treatment of initial conditions), we can use the impulse response to determine the system properties.

- *Static:* An LTI system is static iff [if and only if] its impulse response is of the form $h(t) = K\delta(t)$, where K is a constant. This follows from the sifting property of the unit impulse.
- *Causality:* An LTI system is causal iff $h(t) = 0$ for all $t < 0$. This follows from the limits of the convolution integral.

  Observe that if a system is causal and if the input $x(t) = 0$ for $t < 0$, then the bounds in the convolution integral can be simplified to

$$ZSR(t) = \int_0^t h(t - \tau)x(\tau)d\tau \tag{2.6}$$

- *BIBO Stability:* An LTI system is BIBO stable iff

$$\int_{-\infty}^{\infty} |h(\tau)|d\tau < \infty$$

i.e., iff the IR is absolutely integrable.
The proof of this result is given in Appendix B.1.2.
Using this test, we conclude that the first-order system with IR $h(t) = be^{-at}u(t)$ is BIBO stable iff $a > 0$.

2.3 LTI Systems: Frequency-Domain Concepts

Recall the notions of *eigenvectors* and *eigenvalues* for square matrices: if A is an $n \times n$ matrix with real entries, then the $n \times 1$ vector v and the (possibly complex) number λ form an eigenvector–eigenvalue pair if $Av = \lambda v$. Think of this result as follows: v is an "input" to system "A" and it results in "output" Av that turns out to be the same as the input except for multiplication by scalar λ (which depends on v). This is of course not true for all v's; this is why those v's for which this is true, and the corresponding "gains" (the λ's), are of special interest. (Note that if v is an eigenvector with corresponding eigenvalue λ, then so is any multiple of v.)

2.3.1 Eigenfunctions of LTI Systems

Now, let us think of LTI systems in the same manner as above. We ask the question: Are there "special" inputs that when applied to an LTI system "reproduce" themselves at the output, except possibly for multiplication by some scalar gain (which could be complex)? Since such inputs are functions of time, we will call them *eigenfunctions of LTI systems*.

> **Eigenfunction Theorem:** Complex exponential functions of the form
>
> $$e^{s_0 t}, \quad -\infty < t < \infty$$
>
> where $s_0 = \sigma_0 + j\omega_0$ is a complex number, are *eigenfunctions for all LTI systems*, i.e.,
>
> $$x(t) = e^{s_0 t}, \ -\infty < t < \infty \ \Rightarrow \ y(t) = \lambda_0 e^{s_0 t}, \ -\infty < t < \infty$$
>
> Moreover, their corresponding *eigenvalues*, namely the λ_0's, are given by
>
> $$\lambda_0 = \int_{-\infty}^{\infty} h(t) e^{-s_0 t} dt \overset{\Delta}{=} H(s_0)$$
>
> a complex number that depends on the system's impulse response $h(t)$ and on the *frequency* s_0 of the input; note that $H(s_0)$ does not depend on t.

Proof of Eigenfunction Theorem

$$y(t) = \int_{-\infty}^{\infty} h(\tau) x(t-\tau) d\tau = \int_{-\infty}^{\infty} h(\tau) e^{s_0(t-\tau)} d\tau = e^{s_0 t} \int_{-\infty}^{\infty} h(\tau) e^{-s_0 \tau} d\tau$$

$$= H(s_0) e^{s_0 t}$$

with the above definition for $H(s_0)$. \hfill Q.E.D.

This proves that $e^{s_0 t}$ is indeed an eigenfunction, and since the form of the result holds for any $h(t)$, $e^{s_0 t}$ is an eigenfunction *for all* LTI systems. Moreover, this is true for any choice of s_0. Different values of s_0 result in different $\lambda_0 = H(s_0)$, and of course the value of $H(s_0)$ depends on the particular system under consideration.

For the moment, we assume that the integral $\int_{-\infty}^{\infty} h(t)e^{-s_0 t} dt$ is finite, i.e., $H(s_0)$ exists. Conditions that guarantee this will be discussed in later chapters. Observe that the eigenfunction theorem is a *steady-state* result in the sense that the input is being applied from $-\infty$ to $+\infty$ (called an *eternal input*) and that the initial state at $-\infty$ is assumed to be zero (since we consider only the ZSR).

Recall that the complex exponential functions $e^{j\omega_0 t}$, i.e., $e^{s_0 t}$ when $\sigma_0 = 0$, are functions that correspond to rotating around the unit circle of the complex plane with velocity of $\frac{\omega_0}{2\pi}$ rotations per second. These functions serve to represent $\cos(\omega_0 t)$ and $\sin(\omega_0 t)$ (and many other functions).

By linearity, we obtain that if input

$$x(t) = K_1 e^{s_1 t} + K_2 e^{s_2 t} + \cdots + K_n e^{s_n t}, \quad -\infty < t < \infty$$

is applied to a system with impulse response $h(t)$, then the output will be

$$y(t) = K_1 H(s_1) e^{s_1 t} + K_2 H(s_2) e^{s_2 t} + \cdots + K_n H(s_n) e^{s_n t}$$

for $-\infty < t < \infty$ where

$$H(s_i) = \int_{-\infty}^{\infty} h(t)e^{-s_i t} dt$$

The eigenfunction theorem is important because of this result and because we can use sums of complex exponential functions to represent large classes of *real signals*.

2.3.2   Frequency Response Function and Transfer Function

The result

$$x(t) = K_1 e^{s_1 t} + K_2 e^{s_2 t} + \cdots + K_n e^{s_n t}$$
$$\Rightarrow y(t) = K_1 H(s_1) e^{s_1 t} + K_2 H(s_2) e^{s_2 t} + \cdots + K_n H(s_n) e^{s_n t}$$

shows the importance of the complex numbers $H(s_i)$ that are calculated from $h(t)$ by $H(s_i) = \int_{-\infty}^{\infty} h(t)e^{-s_i t} dt$.

For this reason, we *define* the function of s

$$H(s) \stackrel{\Delta}{=} \int_{-\infty}^{\infty} h(t)e^{-st} dt \tag{2.7}$$

and call it the *Transfer Function* (TF) of the system with impulse response $h(t)$. The Transfer Function captures all of the information in $h(t)$ and its value at a specific complex number $s = s_0$ gives us the complex number $H(s_0)$ that multiplies the exponential $e^{s_0 t}$ as it goes through the system. As we will see later, the TF is the *Laplace transform* of the impulse response.

In the special case where attention is restricted to purely imaginary s's, i.e., $s = 0 + j\omega$, we *define* the function of ω

$$H(j\omega) \stackrel{\Delta}{=} \int_{-\infty}^{\infty} h(t)e^{-j\omega t}dt \tag{2.8}$$

and call it the *Frequency Response Function* (FRF) of the system with impulse response $h(t)$. We will use the notation $H(j\omega)$ for an FRF, even though it is a function of independent variable ω, as a reminder that it comes from setting $s = 0 + j\omega$. As we will see later, the FRF is the *Fourier transform* of the impulse response. The FRF has a "more intuitive" interpretation than the TF in the sense that we are very familiar with signals that have simple expressions in terms of $e^{j\omega t}$ functions: $\cos(\omega t)$ and $\sin(\omega t)$.

Identity System A system that "does nothing" is one for which $h(t) = \delta(t)$, since the unit impulse is the identity element of convolution. In this case, from the sifting property of the unit impulse, $H(s) = 1$ and $H(j\omega) = 1$.

Obtaining TF and FRF from Differential Equation It turns out that the TF and the FRF can be obtained *by inspection* of the LCCDE describing the system. Consider the system

$$a_n \frac{d^n y(t)}{dt^n} + a_{n-1} \frac{d^{n-1} y(t)}{dt^{n-1}} + \cdots + a_1 \frac{dy(t)}{dt} + a_0 y(t) =$$
$$b_m \frac{d^m x(t)}{dt^m} + b_{m-1} \frac{d^{m-1} x(t)}{dt^{m-1}} + \cdots + b_1 \frac{dx(t)}{dt} + b_0 x(t)$$

where we usually normalize $a_n = 1$. If we pick $x(t) = e^{s_0 t}$, then since

$$\frac{d^k x(t)}{dt^k} = s_0^k e^{s_0 t} \quad \text{and} \quad y(t) = H(s_0)e^{s_0 t}$$

we obtain, after replacing s_0 by s for the sake of simplicity,

$$a_n H(s)s^n e^{st} + a_{n-1} H(s)s^{n-1} e^{st} + \cdots + a_1 H(s)se^{st} + a_0 H(s)e^{st} =$$
$$b_m s^m e^{st} + b_{m-1} s^{m-1} e^{st} + \cdots + b_1 se^{st} + b_0 e^{st}$$

After simplification, we obtain the expression

$$H(s) = \frac{b_m s^m + b_{m-1} s^{m-1} + \cdots + b_1 s + b_0}{a_n s^n + a_{n-1} s^{n-1} + \cdots + a_1 s + a_0}$$

A similar result holds for the FRF (replace s by $j\omega$):

$$H(j\omega) = \frac{b_m (j\omega)^m + b_{m-1} (j\omega)^{m-1} + \cdots + b_1 (j\omega) + b_0}{a_n (j\omega)^n + a_{n-1} (j\omega)^{n-1} + \cdots + a_1 (j\omega) + a_0}$$

Thus, if we are given the LCCDE describing the system, we do not have to calculate the impulse response in order to derive the TF and the FRF. Of course, *this is assuming that the TF and the FRF actually exist.* In the case of the TF, this issue will be addressed in Chap. 6 when we discuss existence conditions for the Laplace transform. For the FRF, this is discussed next.

Remarks on the FRF We make the following observations about the FRF:

- The FRF always exists for BIBO stable systems. Since $H(j\omega)$ is a complex number for each ω, we must show that its magnitude is bounded:

$$|H(j\omega)| = \left| \int_{-\infty}^{\infty} h(t) e^{-j\omega t} dt \right| \leq \int_{-\infty}^{\infty} |h(t)| |e^{-j\omega t}| dt \leq \int_{-\infty}^{\infty} |h(t)| dt < \infty$$

where the last $<$ follows by BIBO stability.

Recall the first-order LCCDE

$$\frac{dy(t)}{dt} + ay(t) = bx(t)$$

We obtain, by inspection,

$$H_{1st}(j\omega) = \frac{b}{a + j\omega} \qquad (2.9)$$

The same result would be obtained from the definition

$$H_{1st}(j\omega) = \int_{-\infty}^{\infty} b e^{-at} u(t) e^{-j\omega t} dt = \int_{0}^{\infty} b e^{-at} e^{-j\omega t} dt$$

This integral exists and returns the expected answer as long as $a > 0$, which is the BIBO stability condition.

- Whenever $h(t)$ is *real* and absolutely integrable, we have the following properties:

1. The FRF is *conjugate symmetric*:

$$H(-j\omega) = H^*(j\omega) \tag{2.10}$$

where $*$ denotes complex conjugate. This is immediate from Eq. (2.8).

2. The *magnitude* of the FRF, $|H(j\omega)|$, is an *even* function. To see this, we write

$$|H(j\omega)|^2 = H(j\omega)H^*(j\omega) = H(j\omega)H(-j\omega)$$
$$|H(-j\omega)|^2 = H(-j\omega)H^*(-j\omega) = H(-j\omega)H(j\omega)$$
$$\Rightarrow |H(j\omega)|^2 = |H(-j\omega)|^2$$

3. The *phase* of the FRF, $\arg[H(j\omega)]$, is an *odd* function. This is because

$$|H(j\omega)|e^{-j\,\arg[H(j\omega)]} = H^*(j\omega)$$
$$= H(-j\omega) = |H(-j\omega)|e^{j\,\arg[H(-j\omega)]}$$
$$\Rightarrow -\arg[H(j\omega)] = \arg[H(-j\omega)]$$

2.3.3 Response to Sinusoidal Inputs: Sine-In Sine-Out Law

Let $x(t) = \cos(\omega_0 t), \ -\infty < t < \infty$. Then, from Euler's[5] formula, we can write

$$x(t) = \frac{1}{2}e^{j\omega_0 t} + \frac{1}{2}e^{-j\omega_0 t}$$

where the two complex frequencies s_0 of interest are $0 + j\omega_0$ and $0 - j\omega_0$. In this case, we obtain immediately from the eigenfunction theorem that

$$y(t) = \frac{1}{2}H(j\omega_0)e^{j\omega_0 t} + \frac{1}{2}H(-j\omega_0)e^{-j\omega_0 t}$$

where

$$H(j\omega_0) = \int_{-\infty}^{\infty} h(t)e^{-j\omega_0 t}dt$$

Obviously, since $x(t)$ is real and $h(t)$ is real (for "real-world" systems), then the above expression for $y(t)$ should be transformable into a real expression. This requires two steps.

---

[5] Leonhard Euler was a Swiss mathematician and scientist, 1707–1783.

First, since $h(t)$ is real, then $H(-j\omega) = H^*(j\omega)$ by conjugate symmetry. By writing the complex numbers in complex exponential form, we get

$$H(j\omega_0) = |H(j\omega_0)|e^{j\,\arg[H(j\omega_0)]}$$

$$H(-j\omega_0) = H^*(j\omega_0) = |H(j\omega_0)|e^{-j\,\arg[H(j\omega_0)]}$$

Second, we use this result to write

$$y(t) = \frac{1}{2}|H(j\omega_0)|\left[e^{j\omega_0 t}e^{j\,\arg[H(j\omega_0)]} + e^{-j\omega_0 t}e^{-j\,\arg[H(j\omega_0)]}\right]$$

$$= |H(j\omega_0)|\cos(\omega_0 t + \arg[H(j\omega_0)])$$

A similar result holds for sin functions:

$$x(t) = \sin(\omega_0 t), \quad -\infty < t < \infty$$

$$\Rightarrow y(t) = |H(j\omega_0)|\sin(\omega_0 t + \arg[H(j\omega_0)])$$

More generally, we can write

$$x(t) = \cos(\omega_0 t + \theta), \quad -\infty < t < \infty$$

$$\Rightarrow y(t) = |H(j\omega_0)|\cos(\omega_0 t + \theta + \arg[H(j\omega_0)])$$

and again similarly for sin.

These results are extremely important. We refer to them as the "sine-in sine-out law" of LTI systems.

Sine-In Sine-Out Law of LTI Systems: A sinusoidal input produces a sinusoidal output, with two modifications:

1. The amplitude of the sinusoid is multiplied by the *magnitude of the Frequency Response Function* at the frequency of the sinusoid.
2. The phase of the sinusoid is changed by an amount equal to the *phase of the Frequency Response Function* at the frequency of the sinusoid.

This result justifies the name *Frequency Response Function* for $H(j\omega)$. In fact, FRFs are often determined experimentally by applying sinusoids of different frequencies and observing the change in amplitude and the phase shift.

As a special case of the sine-in sine-out law, a constant input will produce a constant output with change in amplitude equal to $H(0)$, a real number:

$$H(0) = \int_{-\infty}^{\infty} h(t)dt$$

where $H(\cdot)$ stands for either the FRF or the TF.

2.3.4 Response to Sums of Sinusoids

If the input to a system can be *exactly* represented as

$$x(t) = \sum_{k=-\infty}^{\infty} X[k]e^{jk\omega_0 t} \tag{2.11}$$

where some $X[k]$'s could be equal to 0, then by the eigenfunction theorem we can immediately write

$$y(t) = \sum_{k=-\infty}^{\infty} X[k]H(jk\omega_0)e^{jk\omega_0 t}$$

In the next chapter, we will see that *periodic signals* admit such representations, where

$$X[k] = \frac{1}{T_0} \int_0^{T_0} x(t)e^{-jk\omega_0 t}dt$$

Thus once the $X[k]$ coefficients have been calculated, it is straightforward to write the expression of the output of the system.

The representation in Eq. (2.11) is called the *Fourier series expansion* of $x(t)$ and the $X[k]$'s are called the *Fourier series coefficients* of $x(t)$.

2.3.5 Complex Impedances for Circuits and Second-Order Systems

As an application of the concept of FRF, let us determine the FRFs corresponding to circuit elements, when the *input* is the *current* going through the element and the *output* is the *voltage* across the element. Since the input is the current and the output is the voltage, and since we are dealing with the FRF, the resulting FRFs will be called the *complex impedances* of the corresponding circuit elements and denoted by $Z(j\omega)$.

- *Resistor:* $y(t) = Rx(t)$ and thus $H(j\omega) = \frac{R}{1}$. We write

$$Z_R(j\omega) \overset{\Delta}{=} R$$

- *Capacitor:* $x(t) = C\frac{dy(t)}{dt}$ and thus $H(j\omega) = \frac{1}{Cj\omega}$. We write

$$Z_C(j\omega) \overset{\Delta}{=} \frac{1}{j\omega C}$$

- *Inductor:* $y(t) = L\frac{dx(t)}{dt}$ and thus $H(j\omega) \overset{\Delta}{=} \frac{Lj\omega}{1}$. We write

$$Z_L(j\omega) \overset{\Delta}{=} j\omega L$$

When analyzing a circuit for its steady-state behavior under sinusoidal inputs, or simply for writing its FRF with respect to some input and output, we can treat each element as a "resistor" provided that we use the complex impedance $Z(j\omega)$ of that element. Here are some simple examples, where the given expressions for the FRFs can be obtained by using the voltage divider property in circuit theory.

1. The FRF of a series RC circuit with a voltage source input and with $v_C(t)$ as the output is

$$H_{lp}(j\omega) = \frac{1}{1 + (j\omega)RC}$$

which is an instantiation of the generic first-order FRF

$$H_{1st}(j\omega) = \frac{b}{a + j\omega}$$

which we encountered previously. In this case, the magnitude is given by

$$|H_{lp}(j\omega)| = \frac{1}{\sqrt{1 + (\omega RC)^2}}$$

an even function, as expected, and at the frequency $\omega_c = \frac{1}{RC}$, we obtain that

$$|H_{lp}(j\omega_c)| = \frac{1}{\sqrt{2}}$$

2. Taking instead the output to be $v_R(t)$, we get

$$H_{hp}(j\omega) = \frac{(j\omega)RC}{1 + (j\omega)RC}$$

(We have used the subscripts "lp" and "hp" in the above FRFs since they are examples of "lowpass" and "highpass" filters, respectively, as we will see later in Sect. 4.3.)
3. The FRF of a series RLC circuit with voltage source input and with $v_C(t)$ as the output is

$$H_{2nd}(j\omega) = \frac{\omega_n^2}{(j\omega)^2 + 2\zeta\omega_n(j\omega) + \omega_n^2} = \frac{1}{(j\frac{\omega}{\omega_n})^2 + 2\zeta j(\frac{\omega}{\omega_n}) + 1} \qquad (2.12)$$

where we have used the standard notation for the *natural frequency* $\omega_n = \frac{1}{\sqrt{LC}}$ and for the *damping ratio* $\zeta = \frac{R}{2}\sqrt{\frac{C}{L}}$ of an FRF where the denominator has order two.

Second-Order System

It is worth discussing further the generic form for a second-order FRF given in Eq. (2.12). When plotting the magnitude of this FRF, the resulting plot may have not only one maximum at $\omega_{max,1} = 0$, but also, under a certain condition, it can have a second maximum at frequency

$$\omega_{max,2} = \omega_n\sqrt{1 - 2\zeta^2}$$

The condition is

$$\zeta < \frac{1}{\sqrt{2}}$$

and the value at the maximum is

$$|H(j\omega_{max,2})| = \frac{1}{2\zeta\sqrt{1 - \zeta^2}}$$

(This can be verified analytically by taking the derivative of the magnitude and setting it equal to 0.)

When $0 < \zeta < \frac{1}{\sqrt{2}}$, we get what is termed *overshoot in frequency*, as the magnitude does not decrease monotonically to 0, but it exhibits a bump around $\omega_n\sqrt{1 - 2\zeta^2}$. In the limiting case where $\zeta = 0$, we can see that the denominator of $H(j\omega)$ is equal to 0 at $\omega = \omega_n$; this is the phenomenon of *resonance* at the natural frequency. Namely, if we excite the system with a sinusoidal function at its natural frequency and there is no damping, then the output will be arbitrarily large.

Chapter 3
Periodic Signals and Fourier Series

3.1 Fourier Series of Periodic Signals

3.1.1 Exponential Form

Definition: A large class of *periodic signals* $x(t)$ with period T_0 and radian frequency $\omega_0 = \frac{2\pi}{T_0}$ can be *exactly* represented by the harmonic series

$$x(t) = \sum_{k=-\infty}^{\infty} X[k]e^{jk\omega_0 t} \tag{3.1}$$

where $k \in \mathbb{Z}$ and the $X[k]$ coefficients are given by

$$X[k] = \frac{1}{T_0} \int_{t_0}^{t_0+T_0} x(t)e^{-jk\omega_0 t}\,dt = \frac{1}{T_0} \int_{<T_0>} x(t)e^{-jk\omega_0 t}\,dt \tag{3.2}$$

where t_0 can be arbitrarily chosen, i.e., the integral covers exactly *one* period of $x(t)$, which is the meaning of the notation $\int_{<T_0>}$.

The representation (3.1) is called the *exponential Fourier*[1] *series* expansion of $x(t)$ and the $X[k]$ coefficients are the corresponding *Fourier series coefficients*. Note that in general the $X[k]$'s are complex.

When T_0 is the fundamental period of $x(t)$ (as the notation suggests), the frequency ω_0 is the *fundamental frequency* of $x(t)$ and $k\omega_0$ is the kth harmonic.

---

[1] Jean-Baptiste Joseph Fourier was a French mathematician, 1768–1830.

© The Author(s), under exclusive license to Springer Nature Switzerland AG 2022
S. Lafortune, *A Guide to Signals and Systems in Continuous Time*,
https://doi.org/10.1007/978-3-030-93027-1_3

(But one could also use Eqs. (3.1) and (3.2) even if T_0 is not the fundamental period.) Thus the Fourier series gives the expression of $x(t)$ in terms of its *frequency content*, as a weighted sum involving ω_0 and harmonics. The coefficient $X[0] = \frac{1}{T_0} \int_{<T_0>} x(t)dt$ is the average value of $x(t)$ over one period. For this reason, $X[0]$ is often called the *DC component* of $x(t)$.

Existence Conditions It can be shown that a periodic function $x(t)$ that satisfies the three conditions below, known as the Dirichlet[2] conditions, will have a representation of the form (3.1):

1. $x(t)$ is absolutely integrable over one period, i.e., $\int_{<T_0>} |x(t)|dt < \infty$.
2. $x(t)$ has at most a finite number of discontinuities in one period.
3. $x(t)$ has at most a finite number of minima and maxima in one period.

We will assume henceforth that the periodic signals of interest satisfy the Dirichlet conditions. Under these conditions, it can be shown that the "$=$" in (3.1) is true everywhere *except* at the points of discontinuity of $x(t)$ where the Fourier series is equal to $\frac{x(t^+)+x(t^-)}{2}$.

Orthogonal Expansion It can be verified by integration that

$$k \neq n \Rightarrow \int_{<T_0>} e^{jk\omega_0 t} e^{-jn\omega_0 t} dt = \int_{<T_0>} e^{jk\omega_0 t} [e^{jn\omega_0 t}]^* dt = 0 \qquad (3.3)$$

$$k = n \Rightarrow \int_{<T_0>} e^{jk\omega_0 t} e^{-jn\omega_0 t} dt = \int_{<T_0>} e^{jk\omega_0 t} [e^{jn\omega_0 t}]^* dt = T_0 \qquad (3.4)$$

We say that $e^{jk\omega_0 t}$, $k \in \mathbb{Z}$, forms an orthogonal set of functions over any interval of the form $[t_0, t_0 + T_0]$. Thus, the Fourier series is an orthogonal expansion in terms of complex exponential functions.

Using the orthogonality result of Eqs. (3.3) and (3.4), we can justify the claimed expression for the Fourier series coefficients in Eq. (3.2). If we substitute Eq. (3.1) into Eq. (3.2), making sure to use a new index variable for the summation, interchange summation and integration, and then apply orthogonality, we get

$$X[k] = \frac{1}{T_0} \int_{<T_0>} x(t) e^{-jk\omega_0 t} dt = \frac{1}{T_0} \int_{<T_0>} [\sum_{n=-\infty}^{\infty} X[n] e^{jn\omega_0 t}] e^{-jk\omega_0 t} dt$$

$$= \frac{1}{T_0} \sum_{n=-\infty}^{\infty} (X[n] \int_{<T_0>} e^{jn\omega_0 t} e^{-jk\omega_0 t}) dt$$

$$= \frac{1}{T_0} (X[k]T_0 + 0 + 0 + \ldots) = X[k]$$

---

[2] Peter Gustav Lejeune Dirichlet was a German mathematician, 1805–1859.

Frequency Spectra The plots of $|X[k]|$ vs. ω and $\arg[X[k]]$ vs. ω are called the *magnitude spectrum* and *phase spectrum* of $x(t)$, respectively. These spectra are called *line spectra* since they are only defined at discrete values of ω, namely $k\omega_0$ for $k \in \mathbb{Z}$. Signals whose spectra contain only a *finite* number of *non-zero* spectral lines are called *band-limited* signals.

Conjugate Symmetry It is straightforward to verify that if $x(t)$ is *real*, then

- $X[-k] = X^*[k]$.
- $|X[k]| = |X[-k]|$, i.e., the *magnitude* of the Fourier series coefficients is an *even* function.
- $\arg[X[-k]] = -\arg[X[k]]$, i.e., the *phase* of the Fourier series coefficients is an *odd* function.

These results follow from the same steps as the corresponding results for the FRF $H(j\omega)$ in Sect. 2.3.2. For this reason, the frequency spectra are often only plotted for positive frequencies.

3.1.2 Combined Trigonometric Form

If $x(t)$ is real, then the exponential Fourier series in Eq. (3.1) can be transformed into the *combined trigonometric Fourier series*

$$x(t) = X[0] + \sum_{k=1}^{\infty} 2|X[k]| \cos(k\omega_0 t + \arg[X[k]]) \tag{3.5}$$

To obtain this form, we use the properties of the $X[k]$'s for real signals. When $k \neq 0$,

$$X[-k]e^{j(-k)\omega_0 t} + X[k]e^{jk\omega_0 t} = |X[-k]|e^{-jk\omega_0 t + j \arg[X[-k]]}$$
$$+ |X[k]|e^{jk\omega_0 t + j \arg[X[k]]}$$
$$= |X[k]|e^{-jk\omega_0 t - j \arg[X[k]]} + |X[k]|e^{jk\omega_0 t + j \arg[X[k]]}$$
$$= 2|X[k]| \cos(k\omega_0 t + \arg[X[k]])$$

from which we obtain Eq. (3.5).

3.1.3 Trigonometric Form

A variant of the combined trigonometric form is also used for real signals. It is called simply the *trigonometric form* and is as follows:

$$x(t) = X[0] + \sum_{k=1}^{\infty} [B[k] \cos(k\omega_0 t) + A[k] \sin(k\omega_0 t)] \tag{3.6}$$

where the $A[k]$ and $B[k]$ coefficients are real and are related to the $X[k]$ coefficients as follows, for $k \geq 1$ (where $\Re[z]$ denotes the real part and $\Im[z]$ denotes the imaginary part of complex number z):

$$B[k] = 2\Re[X[k]] = \frac{2}{T_0} \int_{<T_0>} x(t) \cos(k\omega_0 t) dt$$

$$A[k] = -2\Im[X[k]] = \frac{2}{T_0} \int_{<T_0>} x(t) \sin(k\omega_0 t) dt$$

$$X[k] = \frac{B[k]}{2} - \frac{jA[k]}{2}$$

The derivation of this form is as follows:

$$x(t) = X[0] + \sum_{k=1}^{\infty} (X[-k]e^{j(-k)\omega_0 t} + X[k]e^{jk\omega_0 t})$$

$$= X[0] + \sum_{k=1}^{\infty} ([X[k]e^{jk\omega_0 t}]^* + X[k]e^{jk\omega_0 t})$$

$$= X[0] + \sum_{k=1}^{\infty} 2\Re[X[k]e^{jk\omega_0 t}]$$

$$= X[0] + \sum_{k=1}^{\infty} 2\Re[(\Re[X[k]] + j\Im[X[k]])(\cos(k\omega_0 t) + j\sin(k\omega_0 t))]$$

$$= X[0] + \sum_{k=1}^{\infty} 2(\Re[X[k]] \cos(k\omega_0 t) - \Im[X[k]] \sin(k\omega_0 t))$$

3.1.4 Calculation of Fourier Series Coefficients

When computing a Fourier series representation of a signal $x(t)$ that satisfies the Dirichlet conditions, it is suggested to proceed as follows:

1. Find the fundamental period T_0 and the fundamental frequency ω_0.
2. Calculate $X[0]$.
3. (a) If $x(t)$ is a *finite* sum of sinusoidal functions, then express each such function as a sum of complex exponentials (using Euler's formula) and identify the Fourier series coefficients by inspection.
   (b) Otherwise, calculate $X[k]$ for $k \neq 0$ in terms of k by solving the integral in Eq. (3.2) and simplifying the result as much as possible; in this regard, it may be useful to separately consider the cases of even and odd k's.

To illustrate case 3.a, consider the following signal:

$$x(t) = 10 + \cos(2\pi t) + \sin(3\pi t)$$

Clearly, $\omega_0 = \pi$ (the GCD of 2π and 3π) and thus $T_0 = 2$. We can use Euler's formula and write $x(t)$ in a manner that explicitly highlights its exponential Fourier series coefficients:

$$-\frac{1}{2j}e^{-j3\pi t} + \frac{1}{2}e^{-j2\pi t} + 10 + \frac{1}{2}e^{j2\pi t} + \frac{1}{2j}e^{j3\pi t}$$

In particular, $X[0] = 10$, $X[1] = 0$, and $X[3] = \frac{1}{2j}$.

Regarding case 3.b, we observe that the common signals considered in the engineering analysis of physical systems satisfy the Dirichlet conditions: half-rectified or fully rectified sinusoidal functions, rectangular waves, triangular waves, sawtooth waves, and so forth. Tables of Fourier series coefficients for such signals are readily available in most textbooks or by searching online and not included herein.

We present the case of a periodic rectangular wave, since it previews the duality between the "rect" and "sinc" functions that is further highlighted in Sect. 4.1.3. (The sinc function is defined in Appendix A.1.)

Let $x_0(t) = \text{rect}(t/D)$ and consider the *periodic extension* of $x_0(t)$ with fundamental period $T_0 > D$, resulting in signal

$$x_r(t) = \sum_{k=-\infty}^{\infty} x_0(t - kT_0) = \sum_{k=-\infty}^{\infty} \text{rect}(\frac{t - kT_0}{D}) \tag{3.7}$$

We may readily calculate $X_r[0] = \frac{D}{T_0}$. Regarding $X_r[k]$ for $k \neq 0$, some more work is needed, and after simplifications we obtain

$$X_r[k] = \frac{1}{k\pi} \sin\left(\frac{k\pi D}{T_0}\right) = \frac{D}{T_0}\text{sinc}\left(\frac{kD}{T_0}\right) \tag{3.8}$$

This tells us that the spectrum of $|X_r[k]|$ will consist of spectral lines inside a (rectified) "sinc envelope." Recall that the zero crossings of the sinc function occur when its argument is an integer. If $\frac{T_0}{D} = 5$ for instance, there will be 4 non-zero spectral lines from $\omega = 0$ to $\omega = 4 \times \frac{2\pi}{T_0}$ until the first zero crossing at $k = 5$, i.e., the fifth spectral line has zero magnitude at $\omega = 5 \times \frac{2\pi}{T_0}$.

Properties of Fourier Series Coefficients The following results are useful when calculating Fourier series coefficients:

• *Amplitude Transformation:* If $y(t) = Ex(t) + F$, then

$$Y[0] = EX[0] + F$$

$$Y[k] = EX[k] \text{ for } k \neq 0$$

- *Time Reversal:* If $y(t) = x^*(-t)$, then $Y[k] = X^*[k]$.
  When $x(t)$ is real, this becomes $Y[k] = X[-k]$.
- *Time Shift:* If $y(t) = x(t - t_d)$, then $Y[k] = X[k]e^{-jk\omega_0 t_d}$.

Their proofs follow directly from Eqs. (3.1) and (3.2).

As an example of the use of these properties, let us transform the periodic rectangular wave of Eq. (3.7) into a standard symmetric *square wave* by setting $D = \frac{T_0}{2}$, adding a DC offset $F = -1/2$, an amplification $E = 2$, and a time shift $t_d = \frac{T_0}{4}$:

$$x_{sw}(t) = -\frac{1}{2} + 2x_r(t - \frac{T_0}{4}) = \begin{cases} 1 & \text{if } kT_0 < t < kT_0 + \frac{T_0}{2} \\ -1 & \text{if } kT_0 + \frac{T_0}{2} < t < (k+1)T_0 \end{cases} \quad k \in \mathbb{Z}$$

$$(3.9)$$

We obtain that $X_{sw}[0] = 0$ and $X_{sw}[k]$ for $k \neq 0$ is given by

$$X_{sw}[k] = \frac{2}{k\pi} \sin(\frac{k\pi}{2})e^{-jk\pi/2} \tag{3.10}$$

Thus,

$$x_{sw}(t) = \sum_{k=-\infty}^{\infty} \frac{2}{k\pi} \sin(\frac{k\pi}{2})e^{-jk\pi/2}e^{jk\omega_0 t} = \sum_{k=1,\text{odd}}^{\infty} \frac{4}{\pi k} \sin(k\omega_0 t) \tag{3.11}$$

where the last equality is obtained after manipulating the expression of $X_{sw}[k]$ by using Euler's formula, considering even and odd values of k separately, and mapping to the trigonometric form of the Fourier series, as described in Sect. 3.1.3. *Hence, a square wave can be expressed as an infinite sum of sinusoidal functions.*

3.1.5 Parseval's Theorem

Periodic signals $x(t)$ that satisfy the Dirichlet conditions will be *power signals.* Recall that for such signals the average power is given by

$$P^{avg} = \frac{1}{T_0} \int_{<T_0>} |x(t)|^2 dt$$

Parseval's Theorem

$$\frac{1}{T_0} \int_{<T_0>} |x(t)|^2 dt = \sum_{k=-\infty}^{\infty} |X[k]|^2 \tag{3.12}$$

The proof can be found in Appendix B.1.3.

Parseval's[3] theorem shows that there are two ways of calculating the average power; one should pick the easier one. More importantly, observe that

$$\frac{1}{T_0} \int_{<T_0>} |X[k]e^{jk\omega_0 t}|^2 dt = \frac{1}{T_0} \int_{<T_0>} |X[k]|^2 e^{jk\omega_0 t} e^{-jk\omega_0 t} dt$$

$$= \frac{1}{T_0} \int_{<T_0>} |X[k]|^2 dt$$

$$= |X[k]|^2$$

which means that $|X[k]|^2$ is *the average power of the kth harmonic of* $x(t)$. Therefore, Parseval's theorem implies that the average power of $x(t)$ is the sum of the average power of each (positive and negative) harmonic in $x(t)$. This is *not* an obvious result as power is not a linear function. For obvious reasons, the plot of $|X[k]|^2$ vs. ω is called the *Power Spectrum* of $x(t)$.

Remark About Trigonometric Form If the real signal $x(t)$ is expanded in terms of the trigonometric form (3.6), then the power in the kth harmonic (all harmonics are positive in this case) is given by

$$\frac{1}{T_0} \int_{<T_0>} [B[k]\cos(k\omega_0 t) + A[k]\sin(k\omega_0 t)]^2 dt = \frac{A^2[k]}{2} + \frac{B^2[k]}{2}$$

$$= 2(\Re[X[k]])^2 + 2(\Im[X[k]])^2 = 2|X[k]|^2 = |X[k]|^2 + |X[-k]|^2$$

(recall that $|X[k]|$ is even) which makes sense since the (positive) kth harmonic of the trigonometric series (i.e., $B[k]\cos(k\omega_0 t) + A[k]\sin(k\omega_0 t)$) corresponds to the positive kth and negative $-k$th harmonics of the complex exponential series (i.e., $X[k]e^{jk\omega_0 t} + X[-k]e^{-jk\omega_0 t}$).

3.1.6 Interpretation and Convergence Issues

Interpretation Fourier series tell us that periodic signals can be expressed as possibly infinite sums of complex exponential functions; in Sect. 3.2, we show how to exploit this result in the analysis of LTI systems subject to periodic inputs, leveraging the eigenfunction property of complex exponential functions for LTI systems. The physical intuition is very important here: from now on, we will focus on the "frequency content" of signals. Fourier series serve as a "stepping stone" to the more general Fourier transform for aperiodic signals. As we will see in Chap. 4, a large class of aperiodic signals can be expressed as a continuous sum, i.e., an integral, of complex exponential functions $e^{j\omega t}$ in the continuous variable $\omega \in \mathbb{R}$.

---

[3] Marc-Antoine Parseval was a French mathematician, 1755–1836.

Convergence Consider a finite approximation to a Fourier series, where the summation stops at $\pm N$:

$$x_N(t) = \sum_{k=-N}^{N} X[k]e^{jk\omega_0 t} \tag{3.13}$$

An important issue is the convergence of this summation to $x(t)$ as $N \to \infty$. Under the Dirichlet conditions, in the limit, the sum will converge pointwise to the correct value of $x(t)$, except at the points of discontinuity. However, the behavior of finite sums near the points of discontinuity may exhibit "ripples", something that is known as the Gibbs[4] phenomenon. This is clearly apparent when plotting partial sums of the Fourier series decomposition of the square wave given in Eq. (3.11). We do not discuss this further; the reader is encouraged to consult a more comprehensive textbook, such as [6].

3.2   Fourier Series and LTI Systems

If a periodic signal is applied to a BIBO-stable LTI system (starting at $t = -\infty$ with the system in zero state), then it is straightforward to obtain the Fourier series of the output from the Fourier series of the input and from the FRF $H(j\omega)$ of the system by applying the eigenfunction theorem:

$$x(t) = \sum_{k=-\infty}^{\infty} X[k]e^{jk\omega_0 t} \Rightarrow y(t) = \sum_{k=-\infty}^{\infty} X[k]H(jk\omega_0)e^{jk\omega_0 t} \tag{3.14}$$

and thus the Fourier coefficients of the output are

$$Y[k] = X[k]H(jk\omega_0) \tag{3.15}$$

Similarly, in the case of the combined trigonometric and trigonometric forms, respectively, we apply the sine-in sine-out law and get that

$$x(t) = X[0] + \sum_{k=1}^{\infty} 2|X[k]| \cos(k\omega_0 t + \arg[X[k]]) \Rightarrow$$

$$y(t) = H(j0)X[0] + \sum_{k=1}^{\infty} 2|X[k]||H(jk\omega_0)| \cos(k\omega_0 t + \arg[X[k]] + \arg[H(jk\omega_0)])$$

---

[4] Josiah Willard Gibbs was an American scientist, 1839–1903.

$$x(t) = X[0] + \sum_{k=1}^{\infty} [B[k] \cos(k\omega_0 t) + A[k] \sin(k\omega_0 t)] \Rightarrow$$

$$y(t) = H(j0)X[0] + \sum_{k=1}^{\infty} [B[k]|H(jk\omega_0)| \cos(k\omega_0 t + \arg[H(jk\omega_0)])$$

$$+ A[k]|H(jk\omega_0)| \sin(k\omega_0 t + \arg[H(jk\omega_0)])]$$

The above results provide a solution to Problem IO formulated in Chap. 1 for periodic inputs, without having to do a convolution integral. Moreover, they tell us about the frequency content of the output. Specifically, Eq. (3.14) tells us that an LTI system *cannot create new* frequencies in the output that are not already present in the input; it can only affect the frequencies that are present in the input. On the other hand, some frequencies can be *removed* if the FRF has magnitude 0 at these frequencies. (A nonlinear system can create new frequencies; for instance, consider the system $y(t) = x^2(t)$.) However, the above equations do not immediately tell us about the variation of $y(t)$ with t; for this, the sums must be evaluated.

Let us revisit the trigonometric Fourier series expansion of the square wave $x_{sw}(t)$ of Eq. (3.9), which is given in Eq. (3.11). Suppose we have a system whose FRF has magnitude close to 1 at frequency ω_0 and close to zero at frequencies $k\omega_0$ for $k > 1$; then, if we apply $x_{sw}(t)$ at the input, the output will be close to a pure sinusoid: $\frac{4}{\pi} \sin(\omega_0 t)$. This is the operation of *lowpass filtering* that is discussed in Sect. 4.3.

We can also partially solve Problem RE and Problem SI using Eq. (3.15), since

$$X[k] = Y[k][H(jk\omega_0)]^{-1} \quad \text{and} \quad H[jk\omega_0] = Y[k][X[k]]^{-1}$$

Specifically, as long as $|H[jk\omega_0]| \neq 0$, we can recover the frequency content of a periodic input from that of the output; those frequencies that are removed by a zero magnitude of the FRF are not recoverable though. Regarding the FRF, we can learn it at discrete multiples of a given fundamental frequency (by applying a periodic input with such fundamental frequency) and repeat that process by varying the fundamental frequency of the input; recall our discussion in Sect. 2.3.3 about how to use the sine-in sine-out law for obtaining $H(j\omega)$ experimentally.

In the next chapter, we extend the approach of this chapter to aperiodic signals. The FRF will still play a central role in determining the frequency content of the output from that of the input. This time, however, the input and output signals will be "represented" by means of their *Fourier transforms* instead of their Fourier series. This is what we call *frequency-domain analysis*.

Chapter 4
Analysis of Stable Systems Using the Fourier Transform

4.1 The Fourier Transform of Continuous-Time Signals

We extend the idea of representing periodic signals in terms of complex exponentials (which are eigenfunctions for all LTI systems) to aperiodic signals. Aperiodic signals contain a continuous set of frequencies as opposed to the discrete set of frequencies ($k\omega_0$) present in a Fourier series. The frequency variable will be denoted by ω, where $\omega \in \mathbb{R}$. In place of a discrete set of coefficients (the $X[k]$'s corresponding to the frequencies $k\omega_0$ contained in periodic signal $x(t)$), we will have a function of the continuous variable ω. For (aperiodic) signal $f(t)$, this function of ω is called the *Fourier Transform* of $f(t)$.

4.1.1 Derivation and Existence Conditions

The Fourier transform of $f(t)$ is denoted by $F(j\omega)$ and it is defined as follows:

$$\mathscr{F}[f(t)] \triangleq F(j\omega) = \int_{-\infty}^{\infty} f(t)e^{-j\omega t}dt \qquad (4.1)$$

Recall our introduction of the FRF in Sect. 2.3.2. Thus the FRF is the Fourier transform of the IR. For notational consistency with the FRF, as well as to emphasize the connection with the Laplace transform (covered in Chap. 6), we will use the notation $F(j\omega)$ for a generic Fourier transform, even though the independent variable is ω.

© The Author(s), under exclusive license to Springer Nature Switzerland AG 2022
S. Lafortune, *A Guide to Signals and Systems in Continuous Time*,
https://doi.org/10.1007/978-3-030-93027-1_4

In the same manner that the $X[k]$ coefficients of a Fourier series carry information about the frequency content of a periodic signal $x(t)$, the Fourier transform carries information about the frequency content of an aperiodic signal $f(t)$ (where $\omega \in \mathbb{R}$ this time). For this reason, we often use the term *spectrum* when referring to the Fourier transform of a signal.

Signal $f(t)$ can be uniquely recovered from its Fourier transform $F(j\omega)$ by doing the following integral:

$$f(t) = \frac{1}{2\pi} \int_{-\infty}^{\infty} F(j\omega)e^{j\omega t} d\omega \triangleq \mathscr{F}^{-1}[F(j\omega)] \qquad (4.2)$$

(Compare this expression to the exponential Fourier series.)

Observe that Eqs. (4.1) and (4.2) are nearly symmetric; the 2π is due to the use of ω instead of $f = \frac{\omega}{2\pi}$.

Rectangular Pulse and Modulation

We state at the outset two very useful results about the Fourier transform that will be used repeatedly.

- If $f(t)$ is a rectangular pulse of amplitude 1 and width D centered at the origin, i.e., using the rect notation,

$$f(t) = \text{rect}(t/D)$$

then we get from Eq. (4.1) that

$$F(j\omega) = \frac{2}{\omega} \sin(D\omega/2) = D \, \text{sinc}(\omega\frac{D}{2\pi})$$

From the properties of the sinc function, $F(j\omega)$ crosses the horizontal axis (i.e., is equal to 0) at $\frac{D\omega}{2\pi} = n$, i.e., at $\omega = n\frac{2\pi}{D}$, $n \neq 0$.

- If $x(t)$ has Fourier transform $X(j\omega)$, then

$$\mathscr{F}[x(t)\cos(\omega_0 t)] = \frac{1}{2}X(j(\omega - \omega_0)) + \frac{1}{2}X(j(\omega + \omega_0)) \qquad (4.3)$$

i.e., the frequency content of $x(t)\cos(\omega_0 t)$ contains two "copies" of the frequency content of $x(t)$, each shifted by amount ω_0, one to the left of the vertical axis and one to the right of the vertical axis. To verify this, use Euler's formula and Eq. (4.1). Multiplication of a signal by $\cos(\omega_0 t)$ is called *modulation* or *mixing*.

Derivation of Transform Equations

The derivation of the above Eqs. (4.1) and (4.2) for $\mathscr{F}$ and $\mathscr{F}^{-1}$ is as follows.

– Start from periodic signal $f_{T_0}(t)$, where T_0 denotes the period, and such that

$$\lim_{T_0 \to \infty} f_{T_0}(t) = f(t)$$

where $f(t)$ is the aperiodic signal of interest. $f_{T_0}(t)$ is the periodic extension of $f(t)$.
– Write the Fourier series of $f_{T_0}(t)$:

$$f_{T_0}(t) = \sum_{k=-\infty}^{\infty} F[k] e^{jk\omega_0 t} \quad \text{where} \quad F[k] = \frac{1}{T_0} \int_{<T_0>} f_{T_0}(t) e^{-jk\omega_0 t} dt$$

– Observe that, as $T_0 \to \infty$, $f_{T_0}(t) \to f(t)$, $\omega_0 \to d\omega$, and $k\omega_0 \to \omega$, a *continuous variable*.
– Define

$$F_{T_0}[k\omega_0] = F[k] T_0 = \frac{F[k] 2\pi}{\omega_0} = \int_{-T_0/2}^{T_0/2} f_{T_0}(t) e^{-jk\omega_0 t} dt$$

Then

$$\lim_{T_0 \to \infty} F_{T_0}[k\omega_0] = \int_{-\infty}^{\infty} f(t) e^{-j\omega t} dt = F(j\omega)$$

– From the Fourier series expansion of $f_{T_0}(t)$, we have that

$$f_{T_0}(t) = \frac{1}{2\pi} \sum_{k=-\infty}^{\infty} F_{T_0}[k\omega_0] e^{jk\omega_0 t} \omega_0$$

and in the limit the sum over the discrete set of frequencies becomes an integral over the continuous set of frequencies, i.e.,

$$\lim_{T_0 \to \infty} f_{T_0}(t) = f(t) = \frac{1}{2\pi} \int_{-\infty}^{\infty} F(j\omega) e^{j\omega t} d\omega$$

In conclusion, we see that the Fourier transform tells us about the frequency content of aperiodic signals; the value $F(j\omega)$ is equal to the limit of the Fourier series coefficient $F[k]$ times T_0 when $k\omega_0$ tends to ω as $T_0 \to \infty$.

Existence Conditions

We state two sets of *sufficient conditions* for $f(t)$ that guarantee that $F(j\omega)$ exists.

- Dirichlet conditions, encountered previously for Fourier series and repeated here:

  1. $f(t)$ is absolutely integrable, i.e., $\int_{-\infty}^{\infty} |f(t)| dt < \infty$;
  2. $f(t)$ has at most a finite number of finite discontinuities in any finite interval;
  3. $f(t)$ has at most a finite number of minima and maxima in any finite interval.

  In this case, $\mathscr{F}^{-1}[F(j\omega)] = f(t)$, except at the points of discontinuity of $f(t)$ where the inverse transform returns $\frac{f(t^+)+f(t^-)}{2}$ (as in the case of Fourier series).

- Square-integrability:

$$\int_{-\infty}^{\infty} |f(t)|^2 dt < \infty$$

  In this case, it can be shown that the error $e(t) = \mathscr{F}^{-1}[F(j\omega)] - f(t)$ has zero energy, i.e., $\int_{-\infty}^{\infty} |e(t)|^2 dt = 0$.

(Note that in general neither set of sufficient conditions implies the other.) In the remainder of this chapter, we will ignore these "recovery errors" and claim that for all practical purposes, *there is a one-to-one relationship between a signal and its Fourier transform (spectrum)*.

Therefore, the Fourier transform of an *energy signal* will exist. It turns out that if we allow *impulses* to appear in the expression of $F(j\omega)$, then many *power signals* of interest to us will also have Fourier transforms. This will allow us to incorporate Fourier series and Fourier transforms in a common framework.

If the signal of interest represents an IR, then its Fourier transform will exist if the system is BIBO stable, as was mentioned previously in Sect. 2.3.2.

4.1.2   Linearity, Frequency Spectra, and Bode Plots

Linearity of $\mathscr{F}$ It is easily seen from the definition of the Fourier transform in Eq. (4.1) that $\mathscr{F}[\cdot]$ is a linear operation on functions.

Magnitude and Phase Spectra Since $F(j\omega)$ is in general a complex function, we will have to plot its magnitude $|F(j\omega)|$ and phase $\arg[F(j\omega)]$ separately, as we did previously for the Fourier series coefficients $F[k]$.

As was demonstrated earlier for the FRF (which is a Fourier transform), if $f(t)$ is real, then:

1. $F(j\omega)$ is conjugate symmetric, i.e., $F^*(j\omega) = F(-j\omega)$;
2. $|F(j\omega)|$ is an even function of ω;
3. $\arg[F(j\omega)]$ is an odd function of ω.

In practice, we want to plot $|F(j\omega)|$ and $\arg[F(j\omega)]$ over a long range of frequencies ω. In this case, however, it is difficult to "see" what happens at smaller frequencies. For this reason, we frequently use a *logarithmic scale* for the ω-axis in the magnitude and phase spectra. Moreover, because

$$\log(|X(j\omega)||H(j\omega)|) = \log(|X(j\omega)|) + \log(|H(j\omega)|)$$

it is also customary to use a logarithmic scale for the vertical axis in the magnitude spectrum.

The most widely used graphical representation of $F(j\omega)$ is the *Bode*[1] *Plot* of $F(j\omega)$. It actually consists of two plots:

1. The magnitude Bode plot, which is the plot of

$$20\log_{10}(|F(j\omega)|) \text{ vs. } \omega$$

for $\omega \geq 0$, where a *logarithmic scale* is used for the ω-axis. The units of $20\log_{10}(|F(\omega)|)$ are called *decibels* and denoted by "dB." Note that the word *gain* is frequently used instead of "magnitude."
2. The phase Bode plot, which is the plot of

$$\arg[F(j\omega)] \text{ vs. } \omega$$

for $\omega \geq 0$, where a *logarithmic scale* is used for the ω-axis.

4.1.3 Transforms of Common Energy Signals

Table 4.1 lists frequently used Fourier transforms along with their existence conditions. These can be obtained by performing the integral (4.1).

Table 4.1 Some important Fourier transforms of energy signals

$f(t)$	$F(j\omega)$
$e^{-at}u(t)$, $\Re[a] > 0$	$\frac{1}{a+j\omega}$
$te^{-at}u(t)$, $\Re[a] > 0$	$\frac{1}{(a+j\omega)^2}$
$t^{n-1}e^{-at}u(t)$, $\Re[a] > 0$	$\frac{(n-1)!}{(a+j\omega)^n}$
$e^{-a|t|}$, $\Re[a] > 0$	$\frac{2a}{a^2+\omega^2}$
$\text{rect}(t/D)$	$\frac{2}{\omega}\sin(\omega\frac{D}{2}) = D\text{sinc}(\omega\frac{D}{2\pi})$
$\frac{1}{t\pi}\sin(tB) = \frac{B}{\pi}\text{sinc}(t\frac{B}{\pi})$	$\text{rect}(\omega/2B)$

---

[1] Hendrik Wade Bode was an American electrical engineer and scientist, 1905-1982.

The duality result between "rect" and "sinc" is especially important; recall that we encountered a corresponding relationship between rect and sinc in Sect. 3.1.4. As we will see later, $H(j\omega) = \text{rect}(\omega/2B)$ represents an Ideal LowPass Filter with *bandwidth* equal to B.

4.1.4 Transforms and Inverse Transforms of Impulses

From the sifting property of $\delta(t)$, we obtain that

$$\mathscr{F}[\delta(t)] = 1 \quad \text{and} \quad \mathscr{F}[\delta(t - t_0)] = e^{-j\omega t_0}$$

Thus an impulse contains all the frequencies, equally weighted; in engineering, such a signal is called *white noise*. An interesting consequence of this result is that, since $\mathscr{F}^{-1}[1] = \delta(t)$, then from Eq. (4.2) for $\mathscr{F}^{-1}$, we get

$$\delta(t) = \frac{1}{2\pi} \int_{-\infty}^{\infty} e^{j\omega t} d\omega \tag{4.4}$$

The result in Eq. (4.4) has to be interpreted in the limit:

$$\delta(t) = \frac{1}{2\pi} \lim_{\alpha \to \infty} \int_{-\alpha}^{\alpha} e^{j\omega t} d\omega = \frac{1}{2\pi} \lim_{\alpha \to \infty} \frac{2}{t} \sin(\alpha t) = \lim_{\alpha \to \infty} \frac{\alpha}{\pi} \text{sinc}(\frac{\alpha}{\pi} t)$$

which does make sense given the graph of the sinc function.

The inverse transform of an impulse in the frequency domain is computed similarly using the definition of $\mathscr{F}^{-1}$:

$$\mathscr{F}^{-1}[\delta(\omega - \omega_0)] = \frac{1}{2\pi} e^{j\omega_0 t} \tag{4.5}$$

In particular,

$$\mathscr{F}[1] = 2\pi \delta(\omega)$$

4.1.5 Transforms of Some Power Signals and of Periodic Signals

Signum Function The signum or "sgn" function defined in Appendix A.2 is a power signal whose Fourier transform exists and moreover that transform does not contain impulses. To obtain it, we view the constant function 1 as the limit of a sequence of energy signals and write, for $\alpha > 0$,

$$\text{sgn}(t) = \lim_{\alpha \to 0} e^{-\alpha|t|} \text{sgn}(t)$$

(Observe that $\lim_{\alpha \to 0} e^{-\alpha|t|} = 1$.) Using this representation, we proceed to calculate the Fourier transform:

$$\mathscr{F}[\text{sgn}(t)] = \lim_{\alpha \to 0} \mathscr{F}[e^{-\alpha|t|}\text{sgn}(t)] = \lim_{\alpha \to 0} [\frac{-1}{\alpha - j\omega} + \frac{1}{\alpha + j\omega}] = \frac{2}{j\omega}$$

We will not consider in this book $\mathscr{F}[u(t)]$, which can be obtained from the transform of the signum function. As we will see in Chap. 6, it is more natural to use the Laplace transform for this power signal.

Sinusoidal Functions Using Euler's formula and Eq. (4.5), it is straightforward to verify that:

$$\mathscr{F}[\cos(\omega_0 t)] = \pi \delta(\omega - \omega_0) + \pi \delta(\omega + \omega_0) \qquad (4.6)$$

$$\mathscr{F}[\sin(\omega_0 t)] = j\pi \delta(\omega + \omega_0) - j\pi \delta(\omega - \omega_0) \qquad (4.7)$$

Thus the Fourier transforms of sinusoidal functions contain impulses. The same is true for any periodic function that possesses a Fourier series representation, as we now see.

Transforms of Fourier Series If

$$f(t) = \sum_{k=-\infty}^{\infty} F[k]e^{jk\omega_0 t}$$

then

$$F(j\omega) = \sum_{k=-\infty}^{\infty} 2\pi F[k]\delta(\omega - k\omega_0)$$

i.e., the Fourier transform of a Fourier series is what we call a "train of impulses."

Of course, one may ask: Why should we compute Fourier transforms of sinusoidal functions or of periodic functions, since we can achieve our goals with Fourier series? The answer is that we do this simply for the sake of uniformity, so that we can treat these signals together with aperiodic ones, in the domain of continuous frequencies.

The transforms in Eqs. (4.6) and (4.7) are special cases of the transform of a Fourier series. Another interesting special case is a *train of impulses* in the time domain.

Train of Impulses Let

$$f(t) = \sum_{k=-\infty}^{\infty} \delta(t - kT_0)$$

Since $f(t)$ is periodic with period T_0, we can write its Fourier series and obtain (by doing the Fourier series coefficient integral):

$$f(t) = \sum_{k=-\infty}^{\infty} \frac{1}{T_0} e^{jk\omega_0 t}$$

from which we get

$$F(j\omega) = \sum_{k=-\infty}^{\infty} \omega_0 \delta(\omega - k\omega_0) \qquad (4.8)$$

Thus the Fourier transform of a train of impulses spaced by amount T_0 in the time domain is a train of impulses spaced by amount ω_0 in the frequency domain! We will make use of this result in Chap. 5 on sampling.

4.2 Properties of the Fourier Transform and Their Applications

4.2.1 Convolution and Modulation Properties

Convolution Property

If all the involved Fourier transforms exist, then

$$x(t) * h(t) \overset{\mathscr{F}}{\leftrightarrow} X(j\omega) H(j\omega) \qquad (4.9)$$

Interpretation Knowing the spectrum of the input, namely $X(j\omega)$, and the FRF $H(j\omega)$ of a BIBO-stable LTI system, it is straightforward to obtain the spectrum of the output, $Y(j\omega)$: only a multiplication suffices. Thus:

Convolution in the time domain becomes multiplication in the frequency domain.

This property is arguably *the* reason why we use transforms! It is the "continuous-frequency" equivalent of $Y[k] = X[k]H(jk\omega_0)$. The proof of this result is based on the definitions of the convolution integral and of the Fourier transform, and on changing the order of two integrals; see Appendix B.2.1.

If we want to recover $y(t)$, then we need to do $\mathscr{F}^{-1}$. For that, we can consult tables and use the technique of *Partial Fraction Expansion* (PFE). We will study

PFE in Sect. 6.2.1 in the context the Laplace transform, but the same general process can be followed for the inverse Fourier transform. Most of the time, our main focus will be the frequency content (or spectrum) of $y(t)$; in this case, we do not need to do $\mathscr{F}^{-1}$.

The result in (4.9) and the one-to-one relationship between a signal and its Fourier transform provide a way to solve Problem IO without doing convolution in the time domain. Equally importantly, it provides a way to solve Problem RE and Problem SI, at least whenever the magnitude of $H(j\omega)$ or of $X(j\omega)$ is not 0, since we can write:

$$X(j\omega) = Y(j\omega)[H(j\omega)]^{-1} \quad \text{and} \quad H(j\omega) = Y(j\omega)[X(j\omega)]^{-1}$$

Recall that BIBO stability is a sufficient condition for $H(j\omega)$ to exist. How should we handle unstable systems then? Fortunately, a similar result holds for unstable systems, provided we use the *Laplace transform*: convolution in the time domain is multiplication in the s-domain; see Sect. 6.1.2.

Modulation Property

If all the involved Fourier transforms exist, then

$$x(t)m(t) \overset{\mathscr{F}}{\leftrightarrow} \frac{1}{2\pi}X(j\omega) * M(j\omega)$$

The proof is also based on changing the order of two integrals; see Appendix B.2.1.

Interpretation We saw an application of this result when we demonstrated (directly from the definition of $\mathscr{F}$, cf. Eq. (4.3)) that

$$\mathscr{F}[x(t)\cos(\omega_0 t)] = \frac{1}{2}X(j(\omega + \omega_0)) + \frac{1}{2}X(j(\omega - \omega_0))$$

Obviously, it is most of the time easier to do a multiplication in the time domain than a convolution in the frequency domain. However, it is the *frequency interpretation* of this result that matters, in particular, when $m(t)$ is a sinusoid; see Sect. 4.3.2 on *Amplitude Modulation* (and demodulation).

4.2.2 Other Properties

The proofs of the results in this section can be found in Appendix B.2.2. Hereafter, we start from signal $f(t)$ with spectrum $\mathscr{F}[f(t)] = F(j\omega)$.

Time Shifting

$$f(t - t_0) \overset{\mathscr{F}}{\leftrightarrow} e^{-j\omega t_0} F(j\omega)$$

Interpretation: Time shifting corresponds to phase shifting in the frequency domain. Note that the phase shift is a *linear function of ω*: $(-t_0)\omega$. You may have seen the expression "Linear Phase" in the description of audio components; this means that there is no *distortion* of the signal, only a *time delay*.

Frequency Shifting

$$e^{j\omega_0 t} f(t) \overset{\mathscr{F}}{\leftrightarrow} F(j(\omega - \omega_0))$$

for any real ω_0. As an illustration of this property, examine the modulation property with $\cos(\omega t)$ presented earlier.

Time Differentiation

If $f(t)$ is continuous and if $f(t)$ and $\frac{df(t)}{dt}$ are both absolutely integrable (and thus satisfy the Dirichlet conditions), then

$$\frac{df(t)}{dt} \overset{\mathscr{F}}{\leftrightarrow} (j\omega)F(j\omega)$$

Interpretation Differentiation *accentuates the high frequency components* of a signal (due to the multiplication by ω). An LTI system with FRF $H(j\omega) = j\omega$ is a *differentiator*. For example, consider an inductor with current as input and voltage as output.

By repeated applications of this property (and assuming that all the derivatives are absolutely integrable), we get

$$\frac{d^n f(t)}{dt^n} \overset{\mathscr{F}}{\leftrightarrow} (j\omega)^n F(j\omega)$$

Time Integration

If $f(t)$ is piecewise continuous and if $f(t)$ and $\int_{-\infty}^{t} f(\tau)d\tau$ are both absolutely integrable (and thus satisfy the Dirichlet conditions), then

$$\int_{-\infty}^{t} f(\tau)d\tau \overset{\mathscr{F}}{\leftrightarrow} \frac{1}{j\omega} F(j\omega)$$

Interpretation Integration *attenuates the high frequency components* of a signal (due to the division by ω). We often say that integration is a "smoothing operation." An LTI system with FRF $H(j\omega) = \frac{1}{j\omega}$ is an *integrator*. For example, consider a capacitor with current as input and voltage as output.

We can do repeated applications of this property provided that all the signals involved are absolutely integrable.

Important Remark It is important to emphasize that one has to be careful about the assumptions when applying the time differentiation and time integration properties. For example,

$$f(t) = \delta(t) \overset{\mathscr{F}}{\leftrightarrow} 1$$

$$\int_{-\infty}^{t} f(\tau)d\tau = u(t) \overset{\mathscr{F}}{\leftrightarrow} ??$$

but $u(t)$ is not absolutely integrable, so we cannot say that its Fourier transform is $\frac{1}{j\omega}$.

Frequency Differentiation

The "dual" of the time differentiation property is that

$$(-jt)^n f(t) \overset{\mathscr{F}}{\leftrightarrow} \frac{d^n F(\omega)}{d\omega^n}$$

which can also be written as

$$t^n f(t) \overset{\mathscr{F}}{\leftrightarrow} j^n \frac{d^n F(\omega)}{d\omega^n}$$

(Here, we dropped the "j" in the argument of $F(\cdot)$, since the differentiation is with respect to ω.) This property is easily proved from the definition of the $\mathscr{F}$ operation (and thus not given in the appendix).

Time Scaling

$$f(\alpha t) \overset{\mathscr{F}}{\leftrightarrow} \frac{1}{|\alpha|} F(\frac{\omega}{\alpha})$$

(Again, we dropped the "j" in the argument of $F(\cdot)$, since the scaling is with respect to ω.)

Interpretation:

- If $|\alpha| > 1$, we obtain that *compression in the time domain corresponds to expansion in the frequency domain.*
- If $0 < |\alpha| < 1$, we obtain that *expansion in the time domain corresponds to compression in the frequency domain.*

These observations are corroborated by examining the "rect—sinc" and "sinc—rect" Fourier transform pairs in Table 4.1.

Duality

With a twist of notation (where $F(t)$ denotes function F where argument ω is replaced by t), we can state this property as follows:

$$F(t) \stackrel{\mathscr{F}}{\leftrightarrow} 2\pi f(-\omega)$$

As two examples of this property, compare $\mathscr{F}[\delta(t - t_0)]$ with $\mathscr{F}[e^{-jt_0 t}]$; or compare $\mathscr{F}[\mathrm{rect}(t/W)]$ with $\mathscr{F}[W\,\mathrm{sinc}(t\frac{W}{2\pi})]$.

4.2.3   Parseval's Theorem for Energy Signals

We saw in Sect. 3.1.5 that the Fourier series coefficients can be used to calculate the average power of a periodic signal or of some of its harmonics. Similar results hold in the case of aperiodic signals that possess Fourier transforms.

The "continuous-frequency" version of Parseval's theorem shows how to calculate the total *energy* in an energy signal $f(t)$ from its Fourier transform $F(j\omega)$. The proof is given in Appendix B.1.3.

Parseval's Theorem

$$\int_{-\infty}^{\infty} |f(t)|^2 dt = \frac{1}{2\pi} \int_{-\infty}^{\infty} |F(j\omega)|^2 d\omega \tag{4.10}$$

We define the quantity

$$\mathscr{E}_f(\omega) \stackrel{\triangle}{=} \frac{1}{2\pi} |F(j\omega)|^2$$

to be the *Energy Spectral Density* (or ESD) of signal $f(t)$. The plot of $\mathscr{E}_f(\omega)$ vs. ω is called the *energy spectrum*. $\mathscr{E}_f(\omega)$ is defined for all $\omega \in \mathbb{R}$. Parseval's theorem tells us that the total energy in $f(t)$ is obtained by integrating $\mathscr{E}_f(\omega)$ from $-\infty$ to ∞.

Suppose that signal $x(t)$ is applied to an LTI system with FRF $H(j\omega)$ resulting in output $y(t)$. Then using the fact that $Y(j\omega) = H(j\omega)X(j\omega)$, we obtain that

$$\mathscr{E}_y(\omega) = |H(j\omega)|^2 \mathscr{E}_x(\omega)$$

Therefore, the quantity $|H(j\omega)|^2$ tells us how the ESD of the input is affected as it goes through the LTI system.

The quantity

$$\int_{\omega_1}^{\omega_2} \mathscr{E}_f(\omega)d\omega$$

corresponds to the energy in the signal $f(t)$ that is contained in the *frequency band* (ω_1, ω_2); hence the name "energy spectral density."

Remark Since we chose to define the ESD for all frequencies from $-\infty$ to ∞, we must include "negative" frequencies in the calculations for real signals. For example, the energy of *real* signal $f(t)$ between the "physical" frequencies 10 and 20 rd/sec is given by

$$\int_{-20}^{-10} \mathscr{E}_f(\omega)d\omega + \int_{10}^{20} \mathscr{E}_f(\omega)d\omega = 2\int_{10}^{20} \mathscr{E}_f(\omega)d\omega$$

4.3 Filtering

Filtering means altering the *frequency content* of a signal by passing it through a suitably designed LTI system called the *filter*. Filtering is done for several reasons, among them:

- reduction or elimination of the *noise* in the signal, e.g., sensor noise in control systems;
- adjustment of the signal in different frequency bands, e.g., in audio: equalizer, bass/treble controls for low/high frequencies, loudness compensation, Dolby noise reduction systems, etc.;
- selection of certain frequency bands in modulation and demodulation techniques, in multiplexing and demultiplexing, etc.;
- shaping the response of the system to signals of various frequencies; e.g., an automobile suspension system can be viewed as a filter subject to inputs corresponding to road imperfections.

An electrical filter can be built from passive (R, L, C) components or from active (op-amps) components. Filtering can also be done by appropriate software routines in digital control systems or by digital signal processing (DSP) chips.

In the context of this chapter, a filter is an FRF $H(j\omega)$.

4.3.1 Ideal Filters

Ideal Filters are filters whose *magnitude* is either *zero* or *one* and whose *phase* is *zero* or, if non-zero, is a *linear function* of ω. The range of ω where the magnitude of the filter is equal to 1 is called the *passband* of the filter, while the range where it is equal to 0 is called the *stopband*.

There are four types of ideal filters, *LowPass* (ILPF), *HighPass* (IHPF), *BandPass* (IBPF), and *BandStop* (IBSF), with obvious definitions based on the passband being:

1. $|\omega| < \omega_{co}$ for an ILPF with *cutoff* frequency, or *bandwidth*, ω_{co};
2. $|\omega| > \omega_{co}$ for an IHPF with cutoff frequency ω_{co};
3. $\omega_1 < |\omega| < \omega_2$ for an IBPF with passband (ω_1, ω_2) in the positive frequency range and $(-\omega_2, -\omega_1)$ in the negative frequency range.
4. all frequencies except for $\omega_1 < |\omega| < \omega_2$ for an IBSF with stopband (ω_1, ω_2) in the positive frequency range and $(-\omega_2, -\omega_1)$ in the negative frequency range.

Remark About Phase The reason for *linear phase* was mentioned earlier in the discussion on the time shifting property of the Fourier transform.

- If a filter is of the form $H(j\omega) = e^{-j\omega t_d}$ in its passband, then the only effect on the signal (in the passband) corresponds to a time delay. Thus such a filter will not introduce distortion (other than the desired filtering function in the stopband). (Recall that $\mathscr{F}[x(t - t_d)] = X(j\omega)e^{-j\omega t_d}$.)
- Observe that since the magnitude of an ideal filter is zero in its stopband, the phase in the stopband is irrelevant.

Realizability of Ideal Filters and Design Considerations Ideal filters such as the lowpass filter cannot be *realized* (for real-time or on-line filtering) because they are not causal; observe that their impulse responses are non-zero for $t < 0$. For the ILPF, this is seen in the last entry in Table 4.1; the other cases can be verified by doing their inverse Fourier transforms.

Even if ideal filters were physically realizable, they would not necessarily be desirable for the following reason: abrupt changes in the frequency domain lead to *overshoot* and *ringing* in the time domain. To see this, examine the impulse and step responses of an ILPF. In practice, a gradual (smooth) transition from passband to stopband is often preferable because it results in a "smoother" response in the time domain (e.g., think of an automobile suspension as a lowpass filter).

Nevertheless, the concept of ideal filters is very useful and instructive in analyzing systems.

4.3.2 DSB-SC AM and Synchronous Demodulation

Modulation is the process of shifting the frequency content of an information signal to a different frequency band for the purpose of transmission of that information signal. The reasons for shifting to a different frequency band have to do with: (1) antenna length requirement; (2) attenuation of signals in the atmosphere; (3) interference from other signals; (4) multiplexing of signals; and so forth.

Probably the simplest method of modulation is *Double-SideBand, Suppressed Carrier sinusoidal amplitude modulation*, abbreviated as DSB-SC. DSB-SC consists of forming the product

$$s(t) = m(t)c(t) = m(t)A_c \cos(\omega_c t)$$

where $m(t)$ is the information signal to be transmitted and $c(t)$ is the (sinusoidal) carrier signal; ω_c is the carrier frequency. For simplicity, we assume that $A_c = 1$ hereafter.

In practice, $m(t)$ is the *modified* information signal, after the *original* information signal, denoted by $m_o(t)$, has been passed through an ILPF (for the purpose of this discussion) with cutoff frequency ω_m, where $\omega_m < \omega_c$. Thus $m(t)$ has a finite *bandwidth* of ω_m, namely the magnitude of its frequency spectrum is equal to 0 for $|\omega| > \omega_m$. We refer to such a signal as being *band-limited*. Moreover, since $m(t)$ has not yet been modulated, we say that it is *baseband*. Note that the information about $m_o(t)$ not present in $m(t)$ is lost and not recoverable.

The spectrum of the transmitted signal $s(t)$ is

$$S(j\omega) = \frac{1}{2}M(j(\omega + \omega_c)) + \frac{1}{2}M(j(\omega - \omega_c))$$

Demodulation is the process of recovering the information signal $m(t)$ [not $m_o(t)$] from the received version of the transmitted signal $s(t)$. The technique of *synchronous (or coherent) demodulation* consists of multiplying the received signal (i.e, $s(t)$ corrupted by some noise, which we neglect hereafter) by another $\cos(\omega_c t + \theta)$, then follow with a lowpass filter. The demodulation is said to be synchronous because it involves multiplying the received $s(t)$ with a sinusoid just like the carrier $c(t)$, resulting in signal $v(t)$.

Assuming zero phase for the local oscillator, i.e., $\theta = 0$, we have that

$$V(j\omega) = \frac{1}{2}S(j(\omega + \omega_c)) + \frac{1}{2}S(j(\omega - \omega_c))$$

$$= \frac{1}{4}M(j(\omega + 2\omega_c)) + \frac{1}{2}M(j\omega) + \frac{1}{4}M(j(\omega - 2\omega_c))$$

Thus it suffices to pass $v(t)$ through an ILPF of gain 2 and of bandwidth anywhere between ω_m and $2\omega_c - \omega_m$ to *exactly* retrieve $m(t)$. This claim is justified by the fact that if we recover $M(j\omega)$ in the frequency domain, it means that we have recovered

$m(t)$ in the time domain, due to the one-to-one relationship between a signal and its Fourier transform.

Note that for DSB-SC and synchronous demodulation to work, we need

$$\omega_c - \omega_m > 0 \quad \text{and} \quad 2\omega_c - \omega_m > \omega_m$$

i.e.,

$$\omega_c > \omega_m$$

4.3.3 Frequency-Division Multiplexing

Another illustration of the use of ideal filters is provided by *Frequency-Division Multiplexing* (or FDM). FDM is a technique that allows the transmission of multiple information signals *simultaneously* over a single medium (e.g., atmosphere, coaxial cable, etc.) by *separating the signals in frequency*.

In FDM, lowpass filters are first used to ensure that the various information signals, denoted by $x_1(t)$, $x_2(t)$, ..., are band-limited. Then the technique of DSB-SC is used for each signal, but with different carrier frequencies, $\omega_{c,1}$, $\omega_{c,2}$, ... These carrier frequencies are chosen so that there is *no overlap* in the shifted spectra of the information signals in the spectrum of the "composite" signal, denoted by $w(t)$:

$$w(t) = x_1(t)\cos(\omega_{c,1}t) + x_2(t)\cos(\omega_{c,2}t) + \dots$$

For simplicity, assume that all the $x_i(t)$ are band-limited at the same frequency ω_m. Then we need any two adjacent carrier frequencies to be $2\omega_m$ apart, i.e.,

$$\omega_{c,i+1} - \omega_{c,i} > 2\omega_m$$

All signals can then be transmitted simultaneously without distortion since they do not overlap in the frequency domain.

In commercial AM radio in the US, for instance, the information signals are band-limited to $\omega_m \leq 5$ KHz, where ω_m is as defined in DSB-SC above. (This explains the poor audio quality of AM radio, given that the human ear can hear up to 20 KHz.) The carrier frequencies (allocated by the Federal Communications Commission) are 10 KHz apart and range from 540 to 1700 KHz. Note that 10 KHz is just enough to avoid overlap of the frequency contents of adjacent information signals.

It is not hard to think of a way to "demultiplex" the composite signal $w(t)$ and recover the information signals $x_1(t)$, $x_2(t)$, ..., using ideal filters.

1. If the desired signal to recover is $x_1(t)$, then an IBPF with passband $(\omega_{c1} - \omega_m, \omega_{c1} + \omega_m)$ (plus the mirror image of that interval for negative frequencies) is used to recover exactly $v_1(t)$ from $w(t)$.
2. Then $v_1(t)$ can be demodulated by the process of synchronous demodulation described earlier.

4.3.4 Realizable Filters

Ideal filters are non-causal since their impulse responses are non-zero for $t < 0$. A *realizable* filter is one that operates in real-time, which means that it must be causal. If we want to build a causal lowpass filter, it will not have the "brickwall" behavior around the cutoff frequency of the ideal filter. Rather, it will have a *transition band* where the gain goes from around unity to as close to zero as possible. It is customary to specify the requirements on a realizable lowpass filter in terms of the passband, transition band, and stopband. In this regard, the cutoff frequency of a realizable lowpass filter, denoted by ω_{cof}, is defined to be the frequency at which the magnitude is

$$|H(j\omega_{cof})| = 0.707 \quad \text{or} \quad 20\log(|H(j\omega_{cof})|) = -3\,\text{dB}$$

This assumes a "normalized" lowpass filter, i.e., one with unity gain (or around unity) in the passband. Since $|H(j\omega_{cof})|^2 = 0.5$, the cutoff frequency is also called the "half-energy" or "half-power" frequency.

There are many classes of filters that are designed in practice. We briefly discuss the class of *Butterworth[2] filters*.

Butterworth Filters A *Butterworth filter* is a filter whose magnitude squared has the form

$$|H(j\omega)|^2 = \frac{1}{1 + (\omega/\omega_{cof})^{2N}} \tag{4.11}$$

where N is the *order* of the filter and ω_{cof} is the desired cutoff frequency.

We make the following observations about Butterworth filters, without proof.

– The Bode plot of the stopband of a Butterworth filter of order N will have a slope of $N \times (-20\,\text{dB/decade})$.
– Butterworth filters exhibit no ripple in the passband; the shape of the magnitude curve is monotonic in the passband (and also in the stopband). For this reason, Butterworth filters are often said to be *maximally flat*.
– The larger N is, the shorter the transition band is.

---

[2] Stephen Butterworth was a British physicist, 1855–1958.

– For all N, the attenuation at ω_{cof} is -3 dB (which of course is consistent with the definition of ω_{cof}!).
– An RC circuit, with $v_C(t)$ as output, is a Butterworth filter of order 1, with $\omega_{cof} = \frac{1}{RC}$.

Once a value is chosen for N, the FRF function $H(j\omega)$ of the filter has to be derived (or extracted) from the above expression (4.11) for $|H(j\omega)|^2$, based on the fact that

$$|H(j\omega)|^2 = H(j\omega)H^*(j\omega) = H(j\omega)H(-j\omega)$$

This can be done: (1) by hand calculations, in order to find the poles of $H(j\omega)$ (not discussed here); (2) using software tools for that purpose; or (3) by consulting appropriate tables in books on filter design. Once the expression of $H(j\omega)$ has been obtained, tables can be consulted to obtain (passive) realizations of these using RLC circuits.

About HighPass and BandPass Filters There are techniques for transforming lowpass filters into: (1) highpass filters by suitably replacing inductors in the design by capacitors, and vice versa; and (2) bandpass filters by nonlinear frequency transformation.

4.3.5 Notions of Bandwidth of Signals and Filters

We already used the term "bandwidth" when specifying the cutoff frequency of an ILPF. Having introduced realizable filters in the preceding section, we can present different notions of bandwidth that are used for a signal or a filter, based on its spectrum. Hereafter, a signal could be an IR, but it does not have to be.

Absolute Bandwidth This notion is used for band-limited signals. Recall that a signal is band-limited if its (magnitude) frequency spectrum is non-zero in a finite interval only.

• A *Baseband signal* (also called lowpass signal) is a signal whose spectrum is centered at zero; in the band-limited case, its spectrum is non-zero in some interval $(-\omega_B, \omega_B)$. In this case, the absolute bandwidth, denoted by BW_{abs}, is defined to be $BW_{abs} = \omega_B$.

  If we convolve in the time domain two baseband band-limited signals with absolute bandwidths ω_A and ω_B, respectively, then the result will be band-limited with absolute bandwidth equal to $\min(\omega_A, \omega_B)$ from the convolution property of Fourier transforms.

  On the other hand, if we multiply in the time domain two baseband band-limited signals with absolute bandwidths ω_A and ω_B, respectively, then the result will be band-limited with absolute bandwidth equal to $\omega_A + \omega_B$ from the modulation property of Fourier transforms and from the resulting convolution integral in the frequency domain.

- A *Bandpass signal* is a band-limited signal whose spectrum is non-zero in some intervals (ω_1, ω_2) and $(-\omega_2, -\omega_1)$. Modulation of a baseband signal results in a bandpass signal. For a bandpass signal, the absolute bandwidth is defined to be $BW_{abs} = \omega_2 - \omega_1$.

Half-Power (or 3-dB) Bandwidth This notion is used for signals that are not necessarily band-limited and whose magnitude spectrum has only one maximum at the center of the bandwidth.

- For a baseband signal $f(t)$ whose spectrum is maximum at $\omega_{max} = 0$, the 3-dB bandwidth, denoted by $BW_{3\,dB}$, is equal to the frequency $\omega_{3\,dB}$ at which

$$\frac{|F(j\omega_{3\,dB})|}{|F(0)|} = \frac{1}{\sqrt{2}}$$

(hence the names "3-dB" or "half-power").
- For a bandpass signal whose spectrum has a maximum occurring at ω_{max} and with 3-dB attenuation on each side of this maximum occurring at ω_1 and ω_2, the 3-dB bandwidth is defined to be $BW_{3dB} = \omega_2 - \omega_1$.

RMS Bandwidth This notion of bandwidth, denoted by BW_{RMS}, is defined for signal $f(t)$ by the equation:

$$[BW_{RMS}]^2 = \frac{\int_{-\infty}^{\infty} \omega^2 |F(j\omega)|^2 d\omega}{\int_{-\infty}^{\infty} |F(j\omega)|^2 d\omega} \tag{4.12}$$

If we define a corresponding notion in the *time domain*, similarly to Eq. (4.12), called *RMS Time Duration*, TD_{RMS},

$$[TD_{RMS}]^2 = \frac{\int_{-\infty}^{\infty} t^2 |f(t)|^2 dt}{\int_{-\infty}^{\infty} |f(t)|^2 dt}$$

then the following important result called *Time-Bandwidth Product* can be shown:

$$BW_{RMS} \times TD_{RMS} \geq \frac{1}{2}$$

The importance of this result, sometimes called an *uncertainty principle*, is that it quantifies the fact that a signal *cannot be both of short frequency bandwidth and of short time duration*. The time scaling property of $\mathscr{F}$ illustrates the same point.

4.4   Key Takeaways

The techniques presented in this chapter allow us to analyze the behavior of systems composed of interconnections of BIBO-stable LTI systems and modulators (also called mixers), together with adders and pure gains, when the input and intermediate signals are defined over the entire time domain from $-\infty$ to ∞ (i.e., eternal signals) and possess Fourier transforms (possibly allowing for impulses in the frequency domain, as considered in Sects. 4.1.4 and 4.1.5).

Most importantly:

- In the frequency domain, the input-output relationship for a BIBO-stable LTI system is obtained by *multiplication*:

$$Y(j\omega) = H(j\omega)X(j\omega)$$

- In the frequency domain, a mixer produces two copies of the spectrum of the input, each shifted by the carrier frequency to the left and to the right:

$$\mathscr{F}[x(t)\cos(\omega_c t)] = \frac{1}{2}X(j(\omega - \omega_c)) + \frac{1}{2}X(j(\omega + \omega_c))$$

The one-to-one relationships between (i) a signal and its spectrum and (ii) an IR and its FRF allow us to work entirely in the frequency domain at no essential loss of information. This is why we can claim that if we recover $M(j\omega)$ in the frequency domain, then it means we have indeed recovered $m(t)$ in the time domain; recall our discussion on synchronous demodulation.

The results in this chapter give us the necessary tools to analyze, in the frequency domain, modulation and demodulation schemes, such as DSB-SC and synchronous demodulation, and multiplexing schemes, such as FDM, that we considered in this chapter. The same knowledge can be used to study related schemes, such as:

- Double-SideBand modulation With Carrier amplitude modulation (DSB-WC), which is used in commercial AM radio as it allows demodulation by *envelope detection*;
- superheterodyning, a technique used in the demodulation stage in commercial AM radio;
- Single SideBand modulation (SSB), which is sometimes used in FDM to reduce the bandwidth required to transmit a signal by half, as compared to DSB;
- Hilbert transforms, a special filter used in SSB schemes and characterized by IR $h_{Hilbert}(t) = \frac{1}{\pi t}$, resulting in $H_{Hilbert}(j\omega) = -j\,\text{sgn}(\omega)$ (this answer can be verified using the duality property);
- Analog Quadrature Amplitude Modulation (QAM), where a mixer using $\sin(\omega_c t)$ is also employed to produce a 90° phase shift.

Chapter 5
Sampling and Reconstruction

5.1 Sampling Theorem

Sampling Theorem: A baseband band-limited signal $x(t)$ with absolute bandwidth $BW_{abs} = \omega_B = 2\pi f_B$ can be *exactly reconstructed* from its samples $x(nT_s)$, $n \in \mathbb{Z}$, where T_s is the sampling interval, provided that the *sampling frequency* $f_s = \frac{1}{T_s}$ is such that

$$f_s > 2f_B \text{ or, equivalently, } \omega_s > 2\omega_B \qquad (5.1)$$

We prefer to use the terminology "sampling frequency" for f_s, which is the number of samples per second, rather than for the resulting angular frequency $\omega_s = 2\pi f_s$. This explains our use of f_s and f_B in the above statement. The frequency $2f_B = \frac{2\omega_B}{2\pi}$ in condition (5.1) is called the *Nyquist*[1] *sampling rate*.

The sampling theorem is a remarkable result, as it tells us that exact reconstruction of a continuous function is possible from a discrete set of sampled values; the assumption that the signal is band-limited is key here, and so is the choice of the sampling frequency. The sampling theorem is often called the Nyquist-Shannon Sampling Theorem, in honor of Nyquist and Shannon.[2]

---

[1] Harry Nyquist was a Swedish-American electrical engineer, 1889–1976.

[2] According to [6], this theorem had been known in the mathematics literature before it first appeared in the communication theory literature in a 1949 paper by Claude Elwood Shannon, an American mathematician and electrical engineer, 1916–2001.

© The Author(s), under exclusive license to Springer Nature Switzerland AG 2022
S. Lafortune, *A Guide to Signals and Systems in Continuous Time*,
https://doi.org/10.1007/978-3-030-93027-1_5

The reconstruction of the signal $x(t)$ from its samples $x(nT_s)$, $n \in \mathbb{Z}$, can be done in one of the two (ideal) ways described below, resulting in reconstructed signal $x_r(t)$, with $x_r(t) = x(t)$ when condition (5.1) holds.

1. Form the special signal

$$x_s(t) = \sum_{n=-\infty}^{\infty} x(nT_s)\delta(t - nT_s) \qquad (5.2)$$

Then pass $x_s(t)$ through an *ideal* lowpass filter of gain T_s and (absolute) bandwidth ω_c, where $\omega_B < \omega_c < \omega_s - \omega_B$. The output of this filter is $x_r(t)$.
2. Use the following *interpolation formula* where $\omega_B < \omega_c < \omega_s - \omega_B$:

$$x_r(t) = \sum_{n=-\infty}^{\infty} x(nT_s)\frac{2\omega_c}{\omega_s}\mathrm{sinc}(\frac{\omega_c(t - nT_s)}{\pi})$$

$$= \sum_{n=-\infty}^{\infty} x(nT_s)\frac{2\omega_c}{\omega_s}\frac{\sin(\omega_c(t - nT_s))}{\omega_c(t - nT_s)} \qquad (5.3)$$

Reconstruction 1 above proceeds as an "on-line" implementation, except that we know that ideal lowpass filters and impulse functions are not realizable. In practice, the on-line reconstruction of $x(t)$ typically uses realizable filters and zero-order holds, as described in Sect. 5.3. Reconstruction 2, based on Eq. (5.3), is often called the Whittaker-Shannon[3] interpolation formula; it is an "offline" calculation that is the time domain equivalent of the ideal filtering operation in reconstruction 1.

The sampling theorem explains, for example, why playback of digital music files works. In the WAV file format, the stored samples are obtained by sampling the audio signal at $f_s = 44.1$ KHz, which is slightly more than twice the 20 KHz maximum frequency that humans can hear. The sampling theorem also explains why we can control *continuous processes*, such as the cruise control system in a car, by embedded processors that work from sampled values.

5.2 Ideal Sampling and Reconstruction

We use the notion of *ideal sampling and reconstruction* to explain and demonstrate the sampling theorem.

Frequency-Domain Analysis Let $x(t)$ be the baseband band-limited ($BW_{abs} = \omega_B$) signal that we wish to sample. Note that $x(t)$ may be the output of a lowpass

---

[3] Edmund Taylor Whittaker was a British mathematician, 1873–1956.

filter to guarantee that it is band-limited; in this context, this lowpass filter is often referred to as an *anti-aliasing filter*.

Ideal sampling consists of multiplying $x(t)$ by the impulse train

$$p(t) = \sum_{n=-\infty}^{\infty} \delta(t - nT_s) \tag{5.4}$$

of period T_s, resulting in signal

$$x_s(t) = x(t)p(t) = \sum_{n=-\infty}^{\infty} x(t)\delta(t - nT_s) = \sum_{n=-\infty}^{\infty} x(nT_s)\delta(t - nT_s)$$

from the sampling property of the unit impulse function.

Signal $x_s(t)$ is often called the *ideal sampling* of $x(t)$. Essentially, each sample value is associated with an impulse function, since we are treating the set of discrete samples over continuous-time.

Next, we want to find $X_s(j\omega) = \mathscr{F}[x_s(t)]$. We claim that

$$X_s(j\omega) = \sum_{n=-\infty}^{\infty} \frac{1}{T_s} X(j(\omega - n\omega_s)) \tag{5.5}$$

where $\omega_s = \frac{2\pi}{T_s}$. This claim can be verified in two ways:

1. Recall Eq. (4.8), the Fourier transform of a train of impulses,

$$P(j\omega) = \frac{2\pi}{T_s} \sum_{n=-\infty}^{\infty} \delta(\omega - n\omega_s)$$

and then use the modulation property of $\mathscr{F}$ to recover expression (5.5).
2. Recall that $p(t)$ can also be written as

$$p(t) = \sum_{n=-\infty}^{\infty} \frac{1}{T_s} e^{jn\omega_s t}$$

by doing its Fourier series expansion. Then use the frequency shifting property of $\mathscr{F}$ to recover expression (5.5).

By plotting $X_s(j\omega)$, the condition (5.1) on the sampling frequency ω_s is immediate if we want to avoid *aliasing*, i.e., overlap of the copies of the spectrum of $x(t)$ in the spectrum of $x_s(t)$. If condition (5.1) is met, then the absence of aliasing allows us to recover *exactly* $x(t)$ by passing $x_s(t)$ through an ILPF of gain T_s and cutoff frequency ω_c satisfying

$$\omega_B < \omega_c < \omega_s - \omega_B$$

This is the *ideal reconstruction* method associated with ideal sampling. We denote the corresponding ILPF by $H_{ideal,recon}(j\omega)$.

The above discussion demonstrates reconstruction 1 described in Sect. 5.1.

Time-Domain Analysis In the time domain, the preceding lowpass filtering operation can be written as

$$x_r(t) = x_s(t) * h(t)$$

where $h(t)$ is the impulse response of the desired ILPF and is given by

$$h(t) = T_s \frac{\omega_c}{\pi} \text{sinc}(\frac{\omega_c t}{\pi})$$

from Table 4.1.

We recover the interpolation formula (5.3) presented in reconstruction 2 in Sect. 5.1 as follows:

$$x_r(t) = [\sum_{n=-\infty}^{\infty} x(nT_s)\delta(t - nT_s)] * [\frac{2\omega_c}{\omega_s} \text{sinc}(\frac{\omega_c t}{\pi})]$$

$$= \sum_{n=-\infty}^{\infty} x(nT_s) \frac{2\omega_c}{\omega_s} \text{sinc}(\frac{\omega_c(t - nT_s)}{\pi})$$

It is important to emphasize that while impulse functions are not practical and ideal lowpass filters are not realizable on-line, the *interpolation formula is true*, since it is an offline result. As long as $\omega_s > 2\omega_B$ and $\omega_B < \omega_c < \omega_s - \omega_B$,

$$x(t) = \sum_{n=-\infty}^{\infty} x(nT_s) \frac{2\omega_c}{\omega_s} \text{sinc}(\frac{\omega_c(t - nT_s)}{\pi})$$

This equation can be simplified if one picks $\omega_c = \frac{1}{2}\omega_s$:

$$x(t) = \sum_{n=-\infty}^{\infty} x(nT_s)\text{sinc}(\frac{1}{T_s}(t - nT_s))$$

5.3 Practical Sampling and Reconstruction

In practice, sampling is done by an Analog-to-Digital Converter (ADC) that obtains a discrete-time signal $x(nT_s)$ from $x(t)$. Ideally, an ADC must act like a switch that is "open" at all times $t \in \mathbb{R}$, except at the time instants nT_s, $n \in \mathbb{Z}$, where it must "close instantaneously" to record sampled value $x(nT_s)$. (One could think of the signal $x_s(t)$ in Eq. (5.2) as the mathematical representation of this process in continuous-time.) In addition, each continuous sampled value $x(nT_s)$ must be quantized and encoded into a set of bits (e.g., 16 bits are used in WAV files for each left/right channel) before it can be processed by digital signal processing hardware. We do not discuss ADCs further and focus instead on reconstruction from sampled values. In addition, we ignore quantization effects in this chapter.

Zero-Order Hold Approximation Let us assume we wish to re-create the original signal $x(t)$ from the stored sampled values $x(nT_s)$, $n \in \mathbb{Z}$ (e.g., a music file stored on a smart phone or streamed to a smart phone). We cannot in practice generate the signal $x_s(t)$ from Eq. (5.2) and then use an ILPF, as described in the ideal reconstruction 1 in Sect. 5.1. Nor can we practically use the time domain equivalent reconstruction 2 either, the interpolation formula (5.3), since it requires using *all* past and future sampled values to reconstruct $x(t)$ in any interval $(nT_s, (n+1)T_s)$, something that cannot be done in a streaming environment, for instance.

Practical reconstruction from the sampled values is often performed by what is called a *Zero-Order Hold* (ZOH) device. This is a *causal* operation that holds the most recent sampled value $x(nT_s)$ constant in the time interval $(nT_s, (n+1)T_s)$. This results in a *staircase approximation* of $x(t)$ that we denote by $x_0(t)$:

$$x_0(t) = \sum_{n=-\infty}^{\infty} x(nT_s)\text{rect}(\frac{t - \frac{T_s}{2} - nT_s}{T_s})$$

Note the time shift by $\frac{T_s}{2}$ since the rect function is centered around the origin.

If only a crude approximation of $x(t)$ is required, then $x_0(t)$ may suffice, in which case reconstruction of $x_0(t)$ from the sampled values $x(nT_s)$ is trivial (in principle). Otherwise, $x_0(t)$ must be suitably filtered to improve the accuracy of the reconstruction. But what filter should be used for this purpose? We first demonstrate that $x(t)$ can be reconstructed *exactly* from $x_0(t)$ if a specific *ideal* filter is allowed. This will tell us what filter to approximate for on-line (or causal) reconstruction.

Reconstruction from ZOH Output If ideal filters are allowed and if the sampling condition (5.1) holds, then we can *exactly reconstruct* $x(t)$ from $x_0(t)$ by passing $x_0(t)$ through the filter $H_{ZOH,recon}(j\omega)$, called the *ZOH reconstruction filter*,

$$H_{ZOH,recon}(j\omega) = \begin{cases} T_s[H_0(j\omega)]^{-1} & \text{if } |\omega| \leq \omega_c \\ 0 & \text{if } |\omega| > \omega_c \end{cases} \qquad (5.6)$$

where $\omega_B < \omega_c < \omega_s - \omega_B$ and where the filter $H_0(j\omega)$ has the following expression:

$$H_0(j\omega) = e^{-j\omega T_s/2} T_s \text{sinc}(\frac{\omega}{\omega_s})$$

To derive this result, we proceed as follows:

1. First, we observe that $x_0(t)$ can be viewed *mathematically* as the result of the convolution $x_s(t) * h_0(t)$, where $x_s(t)$ is the output of the ideal sampler and where $h_0(t)$ is a rectangular pulse of width T_s and height 1, centered at time $\frac{T_s}{2}$:

$$h_0(t) = \text{rect}(\frac{t - \frac{T_s}{2}}{T_s})$$

Indeed,

$$x_s(t) * h_0(t) = \sum_{n=-\infty}^{\infty} x_n(T_s)\delta(t - nT_s) * \text{rect}\left(\frac{t - \frac{T_s}{2}}{T_s}\right)$$

$$= \sum_{n=-\infty}^{\infty} x(nT_s)\text{rect}\left(\frac{t - \frac{T_s}{2} - nT_s}{T_s}\right)$$

2. Second, we recall that $x(t)$ is exactly recovered if $x_s(t)$ is passed through the filter $H_{ideal,recon}(j\omega)$ where $H_{ideal,recon}(j\omega)$ is an ideal lowpass filter of gain T_s and absolute bandwidth ω_c where $\omega_B < \omega_c < \omega_s - \omega_B$.
3. Since $X_0(j\omega) = X_s(j\omega)H_0(j\omega)$, we can write that $X_s(j\omega) = X_0(j\omega)[H_0(j\omega)]^{-1}$. Moreover, $X_s(j\omega)H_{ideal,recon}(j\omega) = X(j\omega)$. Thus, we want the reconstruction filter to be

$$H_{ZOH,recon}(j\omega) = H_{ideal,recon}(j\omega)[H_0(j\omega)]^{-1}$$

so that

$$X_0(j\omega)H_{ZOH,recon}(j\omega) = X(j\omega)$$

as claimed in Eq. (5.6). It is straightforward to calculate $\mathscr{F}[h_0(t)]$ and verify that $H_0(j\omega)$ is as given above. Note that in the passband of $H_{ZOH,recon}(j\omega)$, $|H_0(j\omega)| \neq 0$; hence, the inverse in Eq. (5.6) is well defined.

Thus, a practical DAC will first use a ZOH and then filter the output of the ZOH by a causal approximation of the reconstruction filter in Eq. (5.6). One can verify that increasing the sampling frequency ω_s will result in a more "flat" shape for $H_{ZOH,recon}(j\omega)$ in its passband.

Our motivation for including the more advanced topic of ZOH reconstruction in this introductory book was to highlight how to leverage time-domain and frequency-domain techniques, as studied in this book, to solve non-trivial engineering problems.

Chapter 6
Analysis and Control of Systems Using the Laplace Transform

6.1 The Laplace Transform: What Is It and Why Do We Need It?

Recall from Sect. 2.3 that e^{st}, $s = \sigma + j\omega$, is an eigenfunction for all LTI systems, i.e.,

$$x(t) = e^{st} \quad \Rightarrow \quad y(t) = H(s)e^{st}$$

where the "eigenvalue" $H(s)$ is a complex number that depends on the impulse response $h(t)$ of the system. Specifically, from the proof of the eigenfunction theorem, we know that

$$H(s) = \int_{-\infty}^{\infty} h(t)e^{-st}dt \tag{6.1}$$

The function $H(s)$ is called the *Transfer Function* (TF) of the system and it can be obtained from the impulse response using Eq. (6.1).

We also proved earlier that the transfer function can be obtained directly from the LCCDE describing the system. Namely, if the system is described by

$$a_n \frac{d^n y(t)}{dt^n} + a_{n-1} \frac{d^{n-1}y(t)}{dt^{n-1}} + \cdots + a_1 \frac{dy(t)}{dt} + a_0 y(t) =$$
$$b_m \frac{d^m x(t)}{dt^m} + b_{m-1} \frac{d^{m-1}x(t)}{dt^{m-1}} + \cdots + b_1 \frac{dx(t)}{dt} + b_0 x(t)$$

(where we usually set $a_n = 1$), then

$$H(s) = \frac{b_m s^m + b_{m-1} s^{m-1} + \cdots + b_1 s + b_0}{a_n s^n + a_{n-1} s^{n-1} + \cdots + a_1 s + a_0} \tag{6.2}$$

Finally, recall that the FRF is a special case of the TF when the complex number s is purely imaginary, i.e., $s = j\omega$.

6.1.1 The Two Types of Laplace Transforms

Inspired by Eq. (6.1) and by the Fourier transform techniques presented in Chap. 4, we define two versions of a new transform, called the Laplace[1] transform, for signal $f(t)$.

1. **Bilateral Laplace Transform:** The bilateral Laplace transform of signal $f(t)$, denoted by $F_B(s) = \mathscr{L}_B[f(t)]$, is defined by the integral

$$F_B(s) = \int_{-\infty}^{\infty} f(t) e^{-st} dt \tag{6.3}$$

whenever this integral exists for at least some values of s.

2. **Unilateral Laplace Transform:** The unilateral Laplace transform of signal $f(t)$, denoted by $F(s) = \mathscr{L}[f(t)]$, is defined by the integral

$$F(s) = \int_{0^-}^{\infty} f(t) e^{-st} dt \tag{6.4}$$

whenever this integral exists for at least some values of s. The reason for using 0^- in the definition of $\mathscr{L}$ is to account for the effect of impulses at time $t = 0$. Essentially, we *do* want to integrate these impulses when calculating $F(s)$ and this is why we start the integral at 0^-.

If $f(t) = f(t)u(t)$, then $F_B(s) = F(s)$. Such signals are called *right-sided signals*.

The TF is the bilateral Laplace transform of the impulse response. If the system is causal, then the TF is also the unilateral Laplace transform of the impulse response. In these cases, since $h(t) = h(t)u(t)$, there is no ambiguity when we write $H(s)$, i.e.,

---

[1] Pierre-Simon de Laplace was a French mathematician, 1749–1827.

$$H(s) = TF = \mathscr{L}[h(t)] = \mathscr{L}_B[h(t)]$$

It is easy to verify that both $\mathscr{L}_B$ and $\mathscr{L}$ are *linear* operations. This implies that *the TF of the parallel connection of two systems is the sum of their individual TFs* (as was the case for the IR and the FRF).

6.1.2 Two Key Results

We present at the outset two results that are central in the study of LTI systems using the Laplace transform.

Convolution Property

The key convolution property from Sect. 4.2.1, which allows us to map convolution in the time domain to multiplication in the frequency domain, remains true when the Laplace transform is employed instead of the Fourier transform. The proof is given in Appendix B.3.1.

1. If input $x(t)$ is applied to an LTI system with TF $H_B(s)$, then

$$Y_B(s) = \mathscr{L}_B[y(t)] = \mathscr{L}_B[x(t) * h(t)] = X_B(s)H_B(s)$$

2. If input $x(t) = x(t)u(t)$ is applied to a *causal* LTI system with TF $H(s)$, then

$$Y(s) = \mathscr{L}[y(t)] = \mathscr{L}[x(t) * h(t)] = X(s)H(s)$$

As was mentioned for the corresponding result for the Fourier transform, the above facts not only provide an alternative to the convolution integral for solving Problem IO, but they also provide a way to solve Problem RE and Problem SI. In fact, regarding Problem IO, we will go beyond the ZSR in this chapter, as we will see how to calculate the ZIR using relevant properties of the Laplace transform.

Note also that the convolution property implies that *the TF of the series connection of two systems is the product of their individual TFs*, as was the case for the FRF.

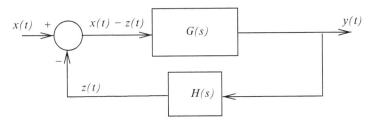

Fig. 6.1 Feedback control architecture with $G(s)$ in forward path and $H(s)$ in negative feedback path

TF of Feedback Loop

We can use linearity and the convolution property to calculate the TF of the standard *negative feedback connection* of two systems, with TF $G(s)$ in the *forward path* and TF $H(s)$ in the *feedback path*, as depicted in Fig. 6.1; this results in the Closed-Loop Transfer Function (CLTF):

$$\text{CLTF} = \frac{Y(s)}{X(s)} = \frac{G(s)}{1 + G(s)H(s)}$$

We will return to the study of feedback control systems in Sect. 6.6, as it is an important application area for Laplace transform techniques.

6.1.3   Why Use the Laplace Transform?

Based on the previous discussion, $\mathscr{L}$ (or $\mathscr{L}_B$) leads to properties that are analogous to those of $\mathscr{F}$. Then why do we bother with $\mathscr{L}$? The answer is that $\mathscr{L}$ *may exist when $\mathscr{F}$ does not exist*. To see this, let us use $\mathscr{L}$ to calculate the output (ZSR) of an LTI system with IR $h(t) = e^{-t}u(t)$ due to input $x(t) = tu(t)$. Note that $\mathscr{F}[x(t)]$ does not exist. It can be shown that $y(t) = [t - 1 + e^{-t}]u(t)$.

In the process, the following $\mathscr{L}$ are obtained from Eq. (6.4):

$$\mathscr{L}[e^{-t}u(t)] = \frac{1}{s+1} \quad \text{and it exists whenever } \Re[s] > -1$$

$$\mathscr{L}[tu(t)] = \frac{1}{s^2} \quad \text{and it exists whenever } \Re[s] > 0$$

$$\mathscr{L}[u(t)] = \frac{1}{s} \quad \text{and exists whenever } \Re[s] > 0$$

Thus we can view $\mathcal{L}$ and $\mathcal{L}_B$ as *generalizations* of $\mathcal{F}$. $\mathcal{L}$ and $\mathcal{L}_B$ are used in the same manner as $\mathcal{F}$ and they have very similar properties. The main differences are that:

1. There are two versions of the Laplace transform, $\mathcal{L}$ and $\mathcal{L}_B$. $\mathcal{L}$ is typically used for the analysis of causal systems subject to right-sided inputs. $\mathcal{L}_B$ is typically used when signals are processed offline (not necessarily in a causal manner).
2. We have to pay attention to the values of s for which $\mathcal{L}_B$ exists, as two different signals may have the same bilateral Laplace transform. For instance, it can be verified from Eq. (6.3) that

$$\mathcal{L}_B[-e^{-t}u(-t)] = \frac{1}{s+1} \quad \text{and it exists whenever } \Re[s] < -1$$

which is the same transform as that of $e^{-t}u(t)$, but with a different existence condition. However, this potential confusion does not arise if we use exclusively the unilateral transform $\mathcal{L}$.

In the remainder of this chapter, we will restrict our attention to the *unilateral* Laplace transform. In other words, we assume that: (1) all systems of interest are causal; and (2) all signals of interest are right-sided. This is motivated by the fact that our primary interest from hereon is the study of the complete response of LCCDEs with initial conditions at time $t_0 = 0$.

6.1.4   Region of Convergence

Definition The set of values of s for which $F(s)$ exists is called the *Region Of Convergence* of $F(s)$, abbreviated as ROC. More precisely, given $f(t) = f(t)u(t)$, writing $s = \sigma + j\omega$ and observing that $|e^{-j\omega t}| = 1$, the ROC is the set of values of s for which

$$\int_{-\infty}^{\infty} |f(t)|e^{-\sigma t}dt < \infty$$

Observe that we can also write

$$F(s) = \int_{-\infty}^{\infty} f(t)e^{-\sigma t}e^{-j\omega t}dt = \mathcal{F}[f(t)e^{-\sigma t}] \tag{6.5}$$

which means that

$$\mathcal{L}[f(t)] \text{ exists} \quad \Leftrightarrow \quad \mathcal{F}[f(t)e^{-\sigma t}] \text{ exists}$$

Roughly speaking, this means that for $F(s)$ to exist, the function $f(t)$ cannot "grow faster" (as $t \to \infty$) than an exponential function. In this case, we can write that $F(s)$ exists whenever

$$f(t) < Me^{\alpha t}$$

It is the value of α that determines the ROC of $f(t)$.

Relation with Fourier Transform From the above discussion, we conclude that:

- *The Fourier transform of $f(t)$ exists iff the ROC of $F(s)$ includes the imaginary axis.*
- Given $f(t) = f(t)u(t)$ and given $F(s) = \mathscr{L}[f(t)]$ whose ROC include the imaginary axis, then

$$F(j\omega) = \mathscr{F}[f(t)] = F(s)|_{s=j\omega}$$

The two earlier examples of $\mathscr{L}[u(t)]$ and $\mathscr{L}[tu(t)]$ illustrate signals that do not have a Fourier transform[2] but that can be handled with the Laplace transform.

6.1.5 Some Important Transforms

Table 6.1 lists frequently used Laplace transforms. These can be obtained by performing the integral (6.4). The ROCs are also listed, for reference; they tell us if the Fourier transforms of the corresponding signals will exist or not.

6.2 Inverse Laplace Transform

Recall Eq. (6.5) for right-sided signal $f(t)$:

$$F(s) = F(\sigma + j\omega) = \int_{-\infty}^{\infty} f(t)e^{-\sigma t}e^{-j\omega t}\,dt = \mathscr{F}[f(t)e^{-\sigma t}]$$

Therefore, we have that

$$f(t)e^{-\sigma t} = \mathscr{F}^{-1}[F(\sigma + j\omega)] = \frac{1}{2\pi}\int_{-\infty}^{\infty} F(\sigma + j\omega)e^{j\omega t}\,d\omega$$

$$\Rightarrow f(t) = \frac{1}{2\pi}\int_{-\infty}^{\infty} F(\sigma + j\omega)e^{(\sigma + j\omega)t}\,d\omega$$

---

[2] If we allow $\delta(\omega)$, then we can write an expression for the Fourier transform of $u(t)$.

Table 6.1 Some important Laplace transforms and their ROCs

$f(t)$	$F(s)$	ROC
$\delta(t)$	1	For all s
$u(t)$	$\frac{1}{s}$	$\Re[s] > 0$
$tu(t)$	$\frac{1}{s^2}$	$\Re[s] > 0$
$e^{-at}u(t)$	$\frac{1}{s+a}$	$\Re[s] > -a$
$t^n e^{-at}u(t)$	$\frac{n!}{(s+a)^{n+1}}$	$\Re[s] > -a$
$\cos(\omega_0 t)u(t)$	$\frac{s}{s^2+\omega_0^2}$	$\Re[s] > 0$
$\sin(\omega_0 t)u(t)$	$\frac{\omega_0}{s^2+\omega_0^2}$	$\Re[s] > 0$
$e^{-at}\cos(\omega_0 t)u(t)$	$\frac{s+a}{(s+a)^2+\omega_0^2}$	$\Re[s] > -a$
$e^{-at}\sin(\omega_0 t)u(t)$	$\frac{\omega_0}{(s+a)^2+\omega_0^2}$	$\Re[s] > -a$
$t\cos(bt)u(t)$	$\frac{s^2-b^2}{(s^2+b^2)^2}$	$\Re[s] > 0$
$t\sin(bt)u(t)$	$\frac{2bs}{(s^2+b^2)^2}$	$\Re[s] > 0$
$\cos^2(bt)u(t)$	$\frac{s^2+2b^2}{s(s^2+4b^2)}$	$\Re[s] > 0$
$\sin^2(bt)u(t)$	$\frac{2b^2}{s(s^2+4b^2)}$	$\Re[s] > 0$
$\frac{1}{2b^3}[\sin(bt)-bt\cos(bt)]u(t)$	$\frac{1}{(s^2+b^2)^2}$	$\Re[s] > 0$

If we do the change of variable $s = \sigma + j\omega$ in this last equality, we obtain the general expression for the inverse Laplace transform:

$$f(t) = \mathcal{L}^{-1}[F(s)] = \frac{1}{2\pi j}\int_{\sigma-j\infty}^{\sigma+j\infty} F(s)e^{st}ds \tag{6.6}$$

The integral (6.6) is called a *contour integral*. The variable σ is a constant that must be in the ROC of the transform. Then the integral is along a straight line parallel to the $j\omega$-axis in the ROC of the transform. The evaluation of this integral is beyond the scope of this book. We make two important remarks:

1. The integral (6.6) clearly shows that the signal $f(t)$ is being expressed as a "weighted sum" of complex exponentials of the form e^{st}.
2. For $F(s)$ that are *rational functions of s*, i.e., the ratio of two polynomials of s, the evaluation of the contour integral (6.6) in some sense "reduces to" doing a *Partial Fraction Expansion* (PFE) of $F(s)$. For this reason, our focus will be on performing inverse Laplace transforms by the PFE method, discussed next.

6.2.1 Partial Fraction Expansion

The technique of Partial Fraction Expansion (PFE) is widely used to calculate *inverse Fourier transforms* and *inverse Laplace transforms*. In this section, we consider inverse Laplace transforms and use the variable s. However, the methodology presented is general and mostly applicable to the case of inverse Fourier transforms, where $s = j\omega$, as long as the more restrictive existence conditions of the Fourier transform are kept in mind. For instance, while $\mathcal{L}^{-1}[\frac{1}{s}] = u(t)$, it is not true that $\mathcal{F}^{-1}[\frac{1}{j\omega}] = u(t)$. On the other hand, from $\mathcal{L}^{-1}[\frac{1}{s+a}] = e^{-at}u(t)$ for $a > 0$, one can indeed write that $\mathcal{F}^{-1}[\frac{1}{j\omega+a}] = e^{-at}u(t)$ since the ROC of this Laplace transform does contain the $j\omega$-axis.

The idea of PFE is to break a transform given in the form of a rational function of s, denoted by $F(s)$, into a sum of terms that have known inverse Laplace transforms (e.g., from available Laplace transform tables). Thus the inverse transforms of these terms are obtained *by inspection*. It is important to observe that *the time signal obtained is only valid for $t \geq 0^+$*. This is because if the signal has a discontinuity at time $t = 0$, the inverse Laplace transform is not "aware" of this discontinuity and it cannot recover the value *before* the discontinuity (i.e., at $t = 0^-$).

Specifics of PFE

We assume that

$$F(s) = \frac{n(s)}{d(s)}$$

where $n(s)$ and $d(s)$ are two polynomials, with degree$[n(s)] = m$, degree$[d(s)] = n$, and $m < n$. In cases where we are given a rational function

$$G(s) = \frac{n_G(s)}{d_G(s)}$$

with degree$[n_G(s)] =$ degree$[d_G(s)]$, then we use long division to write

$$G(s) = K + F(s)$$

where $F(s)$ satisfies the above assumption. Recall that $\mathcal{L}^{-1}[K] = K\delta(t)$.

The typical terms that we try to obtain when decomposing $F(s)$ are the following:

$$\mathcal{L}^{-1}\left[\frac{1}{s+a}\right] = e^{-at}u(t)$$

$$\mathcal{L}^{-1}\left[\frac{1}{(s+a)^k}\right] = \frac{t^{k-1}}{(k-1)!}e^{-at}u(t)$$

$$\mathcal{L}^{-1}\left[\frac{s+a}{(s+a)^2+b^2}\right] = e^{-at}\cos(bt)u(t)$$

$$\mathcal{L}^{-1}\left[\frac{b}{(s+a)^2+b^2}\right] = e^{-at}\sin(bt)u(t)$$

where the case $a = 0$ is also of relevance.

Roots of Polynomials

The first step in the PFE process is to calculate the *roots* of $d(s)$, which are all the (real or complex) numbers p_0 such that

$$d(s)|_{s=p_0} = d(p_0) = 0$$

The roots of $d(s)$ are also called the *poles* of $F(s)$. The roots of $n(s)$ are called the *zeros* of $F(s)$.

Fundamental Theorem of Algebra If degree$[d(s)] = n$, then $d(s)$ has n roots, not necessarily distinct. If we denote these roots as $p_1, p_2, \ldots, p_n$, then we can write

$$d(s) = (s - p_1)(s - p_2)\cdots(s - p_n) = \prod_{i=1}^{n}(s - p_i)$$

Moreover, if the coefficients of $d(s)$ are *real*, then complex roots occur in *complex conjugate pairs*, i.e., if $p_i = \sigma_i + j\omega_i$ is a root, then so if $p_i^* = \sigma_i - j\omega_i$.

Case of Distinct Roots

If the n roots of $d(s)$ are *distinct*, then we have that

$$F(s) = \frac{n(s)}{d(s)} = \sum_{i=1}^{n}\frac{A_i}{s - p_i}$$

where the A_i's are called the *residues* and are given by

$$A_i = (s - p_i)F(s)|_{s=p_i}$$

To see why this is true, observe that we can write

$$F(s) = \frac{n(s)}{(s - p_i)d_{rem}(s)} = \frac{A_i}{s - p_i} + \frac{n_{rem}(s)}{d_{rem}(s)}$$

$$\Rightarrow A_i d_{rem}(s) + n_{rem}(s)(s - p_i) = n(s)$$

$$\Rightarrow A_i d_{rem}(p_i) = n(p_i)$$

$$\Rightarrow A_i = \frac{n(p_i)}{d_{rem}(p_i)} = (s - p_i)F(s)|_{s=p_i}$$

Taking $\mathcal{L}^{-1}$ of each term is immediate from our knowledge base in Table 6.1.

Case of Repeated Roots

Let us consider the case where the roots of $d(s)$ are not all distinct, i.e.,

$$d(s) = (s - p_1)^{r_1} \cdots (s - p_r)^{r_r} = \prod_{i=1}^{r}(s - p_i)^{r_i}$$

where $r < n$ and each $r_i \geq 1$. In this case, we expand $F(s)$ as follows:

$$F(s) = \frac{A_{1,1}}{s - p_1} + \frac{A_{1,2}}{(s - p_1)^2} + \ldots + \frac{A_{1,r_1}}{(s - p_1)^{r_1}}$$

$$+ \ldots$$

$$+ \frac{A_{r,1}}{s - p_r} + \frac{A_{r,2}}{(s - p_r)^2} + \ldots + \frac{A_{r,r_r}}{(s - p_r)^{r_r}}$$

Again, taking $\mathcal{L}^{-1}$ of each term is immediate from our knowledge base in Table 6.1.
It can be shown that the A_{i,r_i-k} coefficients, $0 \leq k \leq r_i - 1$, $1 \leq i \leq r$, can be calculated as follows:

$$A_{i,r_i} = (s - p_i)^{r_i} F(s)|_{s=p_i}$$

$$\ldots$$

$$A_{i,r_i-k} = \frac{1}{k!} \frac{d^k}{ds^k}[(s - p_i)^{r_i} F(s)]|_{s=p_i}$$

$$\ldots$$

$$A_{i,1} = \frac{1}{(r_i - 1)!} \frac{d^{r_i-1}}{ds^{r_i-1}}[(s - p_i)^{r_i} F(s)]|_{s=p_i}$$

The proof is omitted and can be found in many textbooks.

Case of Complex Roots

Consider a real polynomial $d(s)$ with a pair of complex roots of the form $p_i = -\sigma_i + j\omega_i$ and $p_i^* = -\sigma_i - j\omega_i$. By following the "standard procedure," one would get partial fractions of the form

$$\frac{A_i}{s - p_i} + \frac{A_i^*}{s - p_i^*}$$

(where the residues are complex conjugates of one another). But this complex expression would have to be transformed to get the expected real answer. Instead, a common practice is to break $F(s)$ in order to get a term of the form

$$\frac{As + B}{(s - p_i)(s - p_i^*)} = \frac{As + B}{s^2 + 2\sigma_i s + \sigma_i^2 + \omega_i^2} = \frac{As + B}{(s + \sigma_i)^2 + \omega_i^2}$$

Then this term is separated into two terms that correspond to known transforms by writing

$$\frac{As + B}{(s + \sigma_i)^2 + \omega_i^2} = \frac{A(s + \sigma_i)}{(s + \sigma_i)^2 + \omega_i^2} + \frac{(B - A\sigma_i)\omega_i/\omega_i}{(s + \sigma_i)^2 + \omega_i^2}$$

$$= A\mathcal{L}[e^{-\sigma_i t} \cos(\omega_i t) u(t)] + \frac{B - A\sigma_i}{\omega_i} \mathcal{L}[e^{-\sigma_i t} \sin(\omega_i t) u(t)]$$

Observe that the polynomial $(s + \sigma_i)^2 + \omega_i^2$ can be obtained from a general polynomial $s^2 + cs + d$ known to have complex roots (i.e., the discriminant is negative) by *completing the square*:

$$s^2 + cs + d = s^2 + cs + \frac{c^2}{4} - \frac{c^2}{4} + d = (s + \frac{c}{2})^2 + (d - \frac{c^2}{4})$$

The same technique can be used in the case of a pair of *purely imaginary roots*, using this time the transforms of $\cos(bt)u(t)$ and $\sin(bt)u(t)$.

Remark When we have two terms of the form $e^{-\sigma t} \cos(\omega t)u(t)$ and $e^{-\sigma t} \sin(\omega t)u(t)$, these can be combined into a single term of the form

$$e^{-\sigma t} \cos(\omega t + \phi)u(t)$$

by using the trigonometric identity:

$$C\cos(\theta) + D\sin(\theta) = \sqrt{(C^2 + D^2)} \cos(\theta - \arctan(\frac{D}{C})) \tag{6.7}$$

6.3 Properties of the Laplace Transform

The proofs of some of the results in this section can be found in Appendix B.3. Hereafter, we start from signal $f(t) = f(t)u(t)$ with Laplace transform $\mathscr{L}[f(t)] = F(s)$.

Time Differentiation

$$\mathscr{L}[\frac{df(t)}{dt}] = sF(s) - f(0^-) \qquad (6.8)$$

By repeated applications of this property, we get

$$\mathscr{L}[\frac{d^2 f(t)}{dt^2}] = s^2 F(s) - sf(0^-) - f'(0^-)$$

$$\cdots$$

$$\mathscr{L}[\frac{d^n f(t)}{dt^n}] = s^n F(s) - s^{n-1} f(0^-) - s^{n-2} f^{(1)}(0^-) \ldots - f^{(n-1)}(0^-)$$

Time Integration

$$\int_{0^-}^{t} f(\tau)d\tau \overset{\mathscr{L}}{\longleftrightarrow} \frac{1}{s} F(s)$$

Thus a system with TF $H(s) = \frac{1}{s}$ is an *integrator*.
The transforms of $\delta(t)$, $u(t)$, $tu(t)$, etc. illustrate this property.

Frequency Differentiation

The "dual" of the time differentiation property is that

$$(-t)f(t) \overset{\mathscr{L}}{\longleftrightarrow} \frac{dF(s)}{ds}$$

which can be applied repeatedly to yield

$$(-t)^n f(t) \overset{\mathscr{L}}{\longleftrightarrow} \frac{d^n F(s)}{ds^n}$$

The proof is immediate from the definition of $\mathscr{L}$ and omitted.

The transform of $t^k e^{-at} u(t)$ illustrates this property; it can be recovered from that of $e^{-at} u(t)$ by applying the property.

Time Shifting

$$f(t - t_0)u(t - t_0), \quad t_0 > 0 \overset{\mathscr{L}}{\leftrightarrow} e^{-st_0} F(s)$$

The reason for the presence of $u(t - t_0)$ is that we want to make sure that nothing prior to $t = 0$ gets shifted into the range of integration of $\mathscr{L}$.

Frequency Shifting

For any $s_0 \in C$, we have that

$$e^{s_0 t} f(t) \overset{\mathscr{L}}{\leftrightarrow} F(s - s_0)$$

Note that the ROC is affected by the shift; e.g., $\Re[s] > \sigma$ becomes $\Re[s - s_0] > \sigma$ which means $\Re[s] > \sigma + \Re[s_0]$. The proof is immediate from the definition of $\mathscr{L}$ and omitted.

Time Scaling

$$f(\alpha t), \quad \alpha > 0 \overset{\mathscr{L}}{\leftrightarrow} \frac{1}{\alpha} F(\frac{s}{\alpha})$$

The reason for $\alpha > 0$ is to ensure that no values of $f(\cdot)$ prior to $t = 0$ are moved into the range of integration of $\mathscr{L}$. The ROC is affected due to the scaling of s. The proof follows the same steps as the proof of the corresponding property for $\mathscr{F}$.

6.4 Solving Differential Equations Using the Unilateral Laplace Transform

6.4.1 Transfer Function (Revisited) and ZSR

We can use the time differentiation property to recover the earlier result that the transfer function of the (causal) system described by the differential equation, for $t \geq 0$,

$$a_n \frac{d^n y(t)}{dt^n} + a_{n-1}\frac{d^{n-1}y(t)}{dt^{n-1}} + \cdots + a_1\frac{dy(t)}{dt} + a_0 y(t)$$

$$= b_m \frac{d^m x(t)}{dt^m} + b_{m-1}\frac{d^{m-1}x(t)}{dt^{m-1}} + \cdots + b_1\frac{dx(t)}{dt} + b_0 x(t)$$

is

$$H(s) = \frac{b_m s^m + b_{m-1}s^{m-1} + \cdots + b_1 s + b_0}{a_n s^n + a_{n-1}s^{n-1} + \cdots + a_1 s + a_0}$$

To prove this result, recall that the ZSR is given by

$$ZSR(t) = h(t) * x(t)$$

(recall that the convolution integral gives us the ZSR) and from the convolution property

$$\mathscr{L}[ZSR(t)] = H(s)X(s)$$

Since we are dealing with the ZSR, all the IC of the differential equation are zero; then

$$y(0^-) = y^{(1)}(0^-) = \ldots = y^{(n-1)}(0^-) = 0 \Rightarrow \mathscr{L}[y^{(p)}(t)] = s^p Y(s)$$

Moreover, since the input satisfies the property $x(t) = x(t)u(t)$,

$$x(0^-) = x^{(1)}(0^-) = \ldots = x^{(m-1)}(0^-) = 0 \Rightarrow \mathscr{L}[x^{(p)}(t)] = s^p X(s)$$

The above expression of $H(s)$ is then obtained by taking the Laplace transform of both sides of the differential equation.

Therefore, to solve for the ZSR of an LTI system described by a differential equation, we can use the following procedure:

1. Write $H(s)$ by inspection;
2. Calculate $X(s)$ for the given $x(t)$;
3. Calculate $\mathscr{L}^{-1}[H(s)X(s)]$ by PFE.

6.4.2 ZSR and ZIR

One of the main reasons for the usefulness of the Laplace transform is that *it allows us to easily handle non-zero IC*. Let us suppose that we are given an LTI system described by the differential equation, for $t \geq 0$,

$$\frac{d^n y(t)}{dt^n} + a_{n-1}\frac{d^{n-1} y(t)}{dt^{n-1}} + \cdots + a_1\frac{dy(t)}{dt} + a_0 y(t)$$

$$= b_m\frac{d^m x(t)}{dt^m} + b_{m-1}\frac{d^{m-1} x(t)}{dt^{m-1}} + \cdots + b_1\frac{dx(t)}{dt} + b_0 x(t)$$

(note that we set $a_n = 1$ without loss of generality). We wish to calculate the complete system response due to a set of *non-zero* IC *and* to a given input.

To solve this problem, we can use the following procedure:

1. Take $\mathcal{L}$ of both sides of the differential equation, using the differentiation property to incorporate the IC on $y(\cdot)$ and its derivatives;
2. Calculate $X(s)$ for the given input and substitute in the above;
3. Obtain the expression of $Y(s)$ from the above;
4. Obtain $y(t) = ZSR(t) + ZIR(t) = \mathcal{L}^{-1}[Y(s)]$ by PFE.

It is *very instructive* to analyze the general expression of $y(t)$ when we follow the above procedure.

- First we write

$$H(s) = \frac{b_m s^m + b_{m-1}s^{m-1} + \cdots + b_1 s + b_0}{s^n + a_{n-1}s^{n-1} + \cdots + a_1 s + a_0} = \frac{n(s)}{d(s)}$$

- Let us also define

$$p_{ic}(s) = [s^{n-1}y(0^-) + s^{n-2}y^{(1)}(0^-) + \ldots + y^{(n-1)}(0^-)]$$
$$+ a_{n-1}[s^{n-2}y(0^-) + s^{n-3}y^{(1)}(0^-) + \ldots + y^{(n-2)}(0^-)]$$
$$+ \cdots$$
$$+ a_1[y(0^-)]$$

This polynomial captures (up to a change of sign) all the terms involving the IC when we take $\mathcal{L}$ of the LHS of the differential equation.
- Then $Y(s)$ will be of the form

$$Y(s) = \frac{n(s)}{d(s)}X(s) + \frac{p_{ic}(s)}{d(s)} \qquad (6.9)$$

where the first term on the RHS of Eq. (6.9) corresponds to the ZSR and the second term to the ZIR.
- $X(s)$ will itself be of the form

$$X(s) = \frac{n_x(s)}{d_x(s)}$$

Thus we can rewrite Eq. (6.9) as

$$Y(s) = \frac{n(s)n_x(s)}{d(s)d_x(s)} + \frac{p_{ic}(s)}{d(s)} \tag{6.10}$$

Equation (6.10) contains *a considerable amount of information!* When we take $\mathcal{L}^{-1}$ of Eq. (6.10) by PFE, we will get:

1. Terms due to the first term of the RHS of Eq. (6.10)—the ZSR—that are due to the *poles of the TF of the system*, i.e., the roots of denominator polynomial $d(s)$.
   → These are the "transients" in the ZSR.
2. Terms due to the first term of the RHS of Eq. (6.10)—the ZSR—that are due to the *poles of the Laplace transform of the input*, i.e., the roots of denominator polynomial $d_x(s)$.
   → This is the "replica" of the input that appears in the ZSR.
3. Terms due to the second term of the RHS of Eq. (6.10)—the ZIR—that are due to the *poles of the TF of the system*, i.e., the roots of denominator polynomial $d(s)$.
   → These are the "transients" that constitute the ZIR.

Note that the two sets of "transient terms" (in 1 and 3 above) have the same form, as they are both due to the poles of the TF. The time functions that appear in these terms are called the *modes of the system*.

Thus the poles of the TF tell us all that we need to know about the form of the "transients" in the response of the system.

6.5   Pole Locations, Stability, and Time Response

Recall the general expression (6.10) for the complete response of a system with TF $H(s) = \frac{n(s)}{d(s)}$ to a set of IC and to input $x(t)$. The first term in that expression is $\mathcal{L}[ZSR(t)]$ and the second term is $\mathcal{L}[ZIR(t)]$. We now discuss the role of the *poles of $H(s)$* in determining the form of $ZIR(t)$ and of the "transients" in $ZSR(t)$, and in characterizing the *stability* properties (BIBO and asymptotic stability) of the system.

At this point, it is useful to recall the following results:

1. BIBO stability is determined by the impulse response $h(t) = \mathcal{L}^{-1}[H(s)]$; namely, the system is BIBO stable iff

$$\int_{-\infty}^{\infty} |h(t)|dt < \infty$$

2. Asymptotic stability is determined by the ZIR; namely, the system is asymptotically stable if

$$\lim_{t \to \infty} ZIR(t) = 0$$

for all IC.
3. Since we assume that all the coefficients of the TF are real (i.e., we are dealing with a "physical" system), then complex poles of $H(s)$ will occur in complex conjugate pairs.

6.5.1 Pole Locations and Stability

Terminology:

ORHP: Open Right-Half of complex Plane, i.e., $\{s \in \mathbb{C} : \Re[s] > 0\}$.
OLHP: Open Left-Half of complex Plane, i.e., $\{s \in \mathbb{C} : \Re[s] < 0\}$.

Poles in the ORHP

Fact If the TF $H(s)$ has *one or more* poles in the ORHP, then the system is *not* BIBO stable and is *not* asymptotically stable.

Explanation Equation (6.10) shows that in this case, when doing PFE and $\mathscr{L}^{-1}$, we will get a time function of the form

$$e^{at} u(t)$$

if the pole is real and simple, or of the form

$$e^{at} \cos(bt + \phi)u(t)$$

if we have a pair of complex poles in the ORHP, *with $a > 0$ in both $ZSR(t)$ and $ZIR(t)$*.

- With such a $ZIR(t)$, the system is not asymptotically stable.
- With such a term in $ZSR(t)$, by picking $x(t) = u(t)$ (a bounded input), we also conclude that the system is not BIBO stable. Another way to see this is to observe that since the same exponentially increasing term will appear in $h(t)$, then the IR is not absolutely integrable and thus the system is not BIBO stable.

Clearly, if the pole in the ORHP has multiplicity greater than one, then the same conclusions hold.

Poles in the OLHP

Fact If the TF $H(s)$ has *all* of its poles in the OLHP, then the system *is* BIBO stable and *is* asymptotically stable.

Explanation Equation (6.10) shows that in this case, when doing PFE and $\mathcal{L}^{-1}$, then all of the time functions in $ZIR(t)$ and $h(t)$ will be of one of the following forms, with $a > 0$:

1. $e^{-at}u(t)$ for a simple pole at $p = -a$ on the real axis;
2. $t^k e^{-at}u(t)$ with $k \geq 1$ for a pole of multiplicity $k + 1$ at $p = -a$;
3. $e^{-at}\cos(bt + \phi)u(t)$ for a pair of complex poles at $s = -a \pm jb$;
4. $t^k e^{-at}\cos(bt + \phi)u(t)$ with $k \geq 1$ for a pair of complex poles of multiplicity $k + 1$ at $s = -a \pm jb$.

Conclusion:

- Since the $ZIR(t)$ is of this form for all IC, then the system is asymptotically stable.
- Such an $h(t)$ is absolutely integrable, implying BIBO stability.

Poles on the Imaginary Axis

Fact If the TF has *one or more poles* on the imaginary axis, then the system is *not* BIBO stable and is *not* asymptotically stable.

Explanation Such a pole (or poles) will result in one or more of the following terms in both $ZIR(t)$ and $h(t)$:

1. $u(t)$ in the case of a simple pole at the origin;
2. $\cos(bt + \phi)u(t)$ in the case of a simple pair of complex poles on the imaginary axis;
3. $t^k u(t)$ for $k \geq 1$ in the case of multiple poles at the origin;
4. $t^k \cos(bt + \phi)u(t)$ for $k \geq 1$ in the case of multiple pairs of complex poles at the same locations on the imaginary axis.

Conclusion:

- In cases 1 and 2, while $ZIR(t)$ does not grow to ∞, it does not go to 0 and thus the system is not asymptotically stable. These cases are often called cases of *marginal stability* since the ZIR is *bounded*.
- In cases 1 and 2, the IR is not integrable and thus the system is not BIBO stable. Alternatively, pick $x(t) = u(t)$ in case 1 and $x(t) = \cos(bt)u(t)$ in case 2 and verify that the ZSR is unbounded.
- In cases 3 and 4, since both $ZIR(t)$ and $h(t)$ will grow without bound, then the system is neither asymptotically stable nor BIBO stable.

Remark About Cancellations in $H(s)$ So far in our discussion, we have ignored the fact that some factor(s) may cancel out in $n(s)$ and $d(s)$ when obtaining the TF $H(s)$ as described in the derivation leading to Eq. (6.9). In particular, it may happen that $n(s)$ and $d(s)$ have a common unstable factor, as would occur in the LCCDE:

$$\frac{d^2 y(t)}{dt^2} - 9y(t) = \frac{dx(t)}{dt} - 3x(t) \tag{6.11}$$

Strictly speaking, this system *is* BIBO stable, as we are left with a single pole at $p = -3$ after cancelling the common factor $(s - 3)$ when obtaining $H(s)$. However, it is *not* asymptotically stable, since that unstable factor $(s - 3)$ will not cancel out in general between $p_{IC}(s)$ and $d(s)$ in Eq. (6.9); note that asymptotic stability must hold for all IC.

Hence, for the purpose of determining asymptotic or marginal stability, one should use the denominator of the TF *before* any cancellations. If the TF was obtained from an LCCDE, then the denominator before cancellations is the *characteristic polynomial* (or equation) of the LCCDE.

6.5.2 Summary

1. An LTI system is BIBO stable iff all the poles of $H(s)$ are in the OLHP.
2. An LTI system is asymptotically stable iff all the poles of $H(s)$, before cancellations, are in the OLHP.
3. An LTI system is *marginally stable* iff all the poles of $H(s)$, before cancellations, are in the LHP and those on the imaginary axis are simple (i.e., they have multiplicity one in the denominator of $H(s)$).

How to Do the Stability Test Software tools can be used to obtain the poles of a TF, although one should keep in mind that finding roots of a polynomial is numerically difficult especially when these roots have multiplicity larger than one.

As an alternative, or when the TF contains some unassigned parameter(s), one can take advantage of the Routh-Hurwirtz[3] stability test (or criterion) that provides necessary and sufficient conditions for all roots of a polynomial to have negative real parts (i.e., to be in the OLHP). We present, without proofs, instantiations of this test for second and third-order polynomials.

---

[3] Edward John Routh was an English mathematician, 1831–1907. Adolph Hurwitz was a German mathematician, 1859–1919.

1. $s^2 + bs + c$ has all its roots in the OLHP iff $[(b > 0)$ and $(c > 0)]$.
   The quadratic formula can also be used to verify the validity of this condition.
2. $s^3 + a_2 s^2 + a_1 s + a_0$ has all its roots in the OLHP iff $[(a_2 > 0)$ and $(a_0 > 0)$ and $(a_2 a_1 > a_0)]$.

6.5.3 Oscillations in Time Response

We discuss some features of the time response of LTI systems, based on the poles of the TF, i.e., the modes of the system. Based on the preceding section, we conclude that we will see *oscillations* in $ZIR(t)$ and in the "transients" of $ZSR(t)$ iff $H(s)$ has *complex poles*.

- The *frequency* of these oscillations is given by the *imaginary part* of the complex poles, since the time functions will be of the form

$$e^{-at} \cos(bt + \phi)u(t)$$

  for a pair of complex poles at $p = -a \pm jb$.
- The real part of the complex poles determines the exponential envelope of the oscillatory terms.

6.5.4 Final Value Theorem

The Final Value Theorem (FVT) is a result that allows us to calculate $\lim_{t \to \infty} f(t)$ directly from $F(s)$ without having to calculate $\mathcal{L}^{-1}[F(s)]$. Not surprisingly, this only makes sense if the signal $f(t)$ has a limit as $t \to \infty$. Thus the *poles of $F(s)$ can only be in the OLHP with in addition at most one pole allowed at $s = 0$*.

Final Value Theorem Under the condition that all the poles of $sF(s)$ are in the OLHP, we have that

$$\lim_{t \to \infty} f(t) = \lim_{s \to 0} sF(s) \tag{6.12}$$

Observe that under the assumption in the statement of FVT, the ROC of $sF(s)$ contains $s = 0$. The proof of the FVT is given in Appendix B.3.3.

6.5.5 Time Response of Second-Order Systems

The standard form of a second-order stable TF is:

$$H(s) = \frac{\omega_n^2}{s^2 + 2\zeta\omega_n s + \omega_n^2}$$

where ω_n is the natural frequency and ζ is the damping ratio. To guarantee stability, we require that $\omega_n > 0$ and $\zeta > 0$. We considered the corresponding FRF in Eq. (2.12).

From the quadratic formula, we obtain that the poles of $H(s)$ are given by

$$p_{1,2} = -\zeta\omega_n \pm \omega_n\sqrt{\zeta^2 - 1}$$

and they will be complex iff $\zeta < 1$.

Consider now the impulse response $h(t)$ and the step response $y_{step}(t)$ of this TF for different values of ζ. We make the following observations. They all follow from the modes of the system for the stated values of ζ. (Verify by doing $\mathcal{L}^{-1}$ by PFE.)

- The case $0 < \zeta < 1$ is called the *underdamped case*.

  In this case, the IR and the step response have a (damped) oscillatory behavior since the poles are complex. In the case of the step response, this leads to what is termed *overshoot* in the response, as there are intervals of time where the output exceeds its final value.

  - Note that the frequency of the oscillations is equal to $\omega_n\sqrt{1 - \zeta^2}$.
  - This is *not* the same as the frequency $\omega_n\sqrt{1 - 2\zeta^2}$ at which the corresponding FRF achieves its maximum; also, recall that overshoot *in frequency* occurs when $\zeta < 0.707$.

- The case $\zeta > 1$ is called the *overdamped case*.

  In this case, since the poles are real, then both modes are decaying exponentials and the IR and the step response do not exhibit oscillations. The greater ζ is, the slower the step response is to converge to 1.

- The case $\zeta = 1$ is called the *critically damped case*.

  In this case, we have a real pole with multiplicity two. Thus there are no oscillations in the IR and in the step response. Note that the step response has "the fastest response" possible without overshoot.

6.6 Feedback Control: A Brief Introduction

We now revisit feedback control systems, first mentioned in Sect. 6.1.2, in the context of the general negative feedback block diagram in Fig. 6.1 with TF $G(s)$ in the *forward path* and TF $H(s)$ in the *feedback path*, resulting in

$$\text{CLTF} = \frac{Y(s)}{X(s)} = \frac{G(s)}{1 + G(s)H(s)}$$

In this section, "stability" refers to BIBO stability.

6.6.1 *Proportional Feedback*

First, let $G(s) = P(s)$, where $P(s)$ is the TF of a given LTI system, or *plant* in control engineering terminology. We assume that $P(s) = \frac{n_P(s)}{d_P(s)}$, a rational function as would be obtained from an LCCDE description of the plant. Let us use a real scalar gain K in the feedback path, i.e., $H(s) = K$; this is an instance of *proportional feedback*. In this case,

$$\text{CLTF} = \frac{n_P(s)}{d_P(s) + Kn_P(s)}$$

We can see that while the zeros of CLTF are the same as those of $P(s)$, the poles have changed and they are now given by the roots of the polynomial

$$d_P(s) + Kn_P(s)$$

Clearly, $K = 0$ means that there is no feedback and thus we recover $d_P(s)$. As K varies, the poles of CLTF will move from the original poles of $P(s)$ to new locations in the complex plane. The location of the poles of CLTF as a function of gain K is called the *root locus* of open-loop TF $P(s)$.

As a simple example, if $P(s) = \frac{1}{s-p}$, where $p > 0$, then the denominator of CLTF becomes $s - p + K$, resulting in a pole of CLTF at $p_{cl} = p - K$. Thus, while $P(s)$ is unstable, CLTF will be stable whenever $K > p$. Various examples of $P(s)$ can be constructed to illustrate that the use of proportional feedback can: (1) stabilize an unstable open-loop system; (2) destabilize a stable open-loop system; (3) change the damping properties of a second-order sub-system; or (4) never result in a stable CLTF if $P(s)$ is unstable.

Open-Loop Control   Stabilization by feedback is preferable to series connection with a compensator that cancels out an unstable pole. For example, if we connect in series $P(s) = \frac{1}{(s+2)(s-3)}$ with $C_{ol}(s) = (s - 3)$, then the open-loop TF, denoted by OLTF, will be

$$OLTF = \frac{1}{s+2}$$

which is stable. However, if the plant parameter 3 were to change slightly to 3.01, then the cancellation would no longer occur and the unstable pole at $s - 3$ would be present in OLTF:

$$OLTF = \frac{(s-3)}{(s+2)(s-3.01)}$$

This problem would not occur if proportional feedback is used, as the reader can verify.

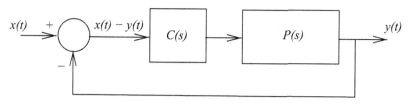

Fig. 6.2 Feedback control architecture with compensator $C(s)$ for plant $P(s)$

6.6.2 PID Compensators

The most common architecture in feedback control is to insert a *compensator* $C(s)$ in series before plant $P(s)$, resulting in $G(s) = C(s)P(s)$ in the forward path, and to set $H(s) = 1$ in the feedback path; refer to Fig. 6.2.

The interpretation here is that a sensor is used to measure the output $y(t)$ of the system (e.g., the velocity of a car in a cruise control system) and the resulting value is subtracted from the input signal $x(t)$ (e.g., the desired speed setting), resulting in an error signal denoted by $e(t) = x(t) - y(t)$ that is used to obtain the actual input for the plant's actuator (e.g., the car's throttle). Specifically, the error signal $e(t)$ is first "filtered" (in the terminology of Chap. 4) by an LTI system with rational TF $C(s) = \frac{n_C(s)}{d_C(s)}$, the compensator, before being sent to the actuator. If we compute the overall CLTF from input $x(t)$ to output $y(t)$, denoted by CLTF-C, then we obtain

$$\text{CLTF-C} = \frac{n_C(s)n_P(s)}{d_C(s)d_P(s) + n_C(s)n_P(s)}$$

A common type of compensator $C(s)$ is the so-called Proportional Integral Derivative (PID) compensator, with TF

$$C(s) = K_P + K_I \frac{1}{s} + K_D s$$

where the three real gains represent: (1) the proportional gain K_P; (2) the integral gain K_I, since an integrator has TF $\frac{1}{s}$; and (3) the derivative gain K_D, since a differentiator has TF s. We can write

$$C(s) = \frac{K_D s^2 + K_P s + K_I}{s}$$

If $K_I = K_D = 0$, then $C(s) = K_P$ and we have *proportional feedback*; the resulting CLTF, denoted by CLTF-P, is

$$\text{CLTF-P} = \frac{K_P n_P(s)}{d_P(s) + K_P n_P(s)}$$

This is the same denominator as in the previous discussion on the root locus of $P(s)$, while the numerator has changed to include gain K_P. Thus the value of K_P will affect the stability properties in the same manner as the case of feedback gain K discussed in Sect. 6.6.1. If this is inadequate, then one may try to use non-zero gains K_I and/or K_D. A thorough discussion of PID compensators is typically covered in the first undergraduate controls course.

6.6.3 Pole Placement Using PD Compensator

As a simple example, consider the second-order plant $P(s) = \frac{1}{s^2-bs-c}$, which is unstable if either $b > 0$ or $c > 0$. Let us use PD compensator $C(s) = \frac{K_D s + K_P}{1}$. The resulting CLTF, denoted by CLTF-PD, is

$$\text{CLTF-PD} = \frac{(K_D s + K_P) n_P(s)}{d_P(s) + (K_D s + K_P) n_P(s)} = \frac{(K_D s + K_P)}{s^2 + (K_D - b)s + (K_P - c)}$$

Interestingly, in this case the poles of CLTF-PD can be placed *anywhere* in the complex plane by proper choice of the two parameters K_D and K_P! This shows the power of feedback in altering the behavior of a system. If we desire a stable system, with critically damped time response corresponding to two poles at $p = -2$, then we solve the equation

$$(s + 2)^2 = s^2 + 4s + 4 = s^2 + (K_D - b)s + (K_P - c)$$

and obtain the gains $K_D = b + 4$ and $K_P = c + 4$.

6.6.4 Zero Steady-State Tracking Error Using PI Compensator

Consider a stable second-order plant $P(s) = \frac{1}{s^2+bs+c}$, where $b > 0$ and $c > 0$. If we use a PI compensator, i.e., $C(s) = \frac{K_P s + K_I}{s}$, then we obtain

$$\text{CLTF-PI} = \frac{(K_P s + K_I) n_P(s)}{s d_P(s) + (K_P s + K_I) n_P(s)} = \frac{(K_P s + K_I)}{s^3 + bs^2 + (c + K_P)s + K_I}$$

Let us assume that K_P and K_I are chosen so that CLTF-PI is stable, which will be the case if

$$0 < K_I < b(c + K_P)$$

from the Routh-Hurwitz criterion for third-order systems.

Assume that we apply a step input $x(t) = Au(t)$ to CLTF-PI. We are interested in the steady-state value of $y(t)$, namely $y_{final} = \lim_{t \to \infty} y(t)$. By stability, we can use the FVT and obtain that

$$y_{final} = \frac{A(K_Ps + K_I)}{s^3 + bs^2 + (c + K_P)s + K_I}\Big|_{s=0} = \frac{AK_I}{K_I} = A$$

Thus, the steady-state error is zero, since $y_{final} = A$. Note that if no compensator is used, then the open-loop steady-state value of the output due to input $Au(t)$ will be

$$\frac{A}{s^2 + bs + c}\Big|_{s=0} = \frac{A}{c}$$

resulting in a steady-state error of $\frac{A}{c} - A = A\frac{(1-c)}{c}$, or 100% error if $c = \frac{1}{2}$ for instance. Thus, one of the advantages of PI compensators is their potential to eliminate the steady-state error.

6.7 Key Takeaways

Recall our concluding remarks in Sect. 4.4 regarding the analysis of BIBO-stable LTI systems in the frequency domain, by leveraging the Fourier transform and its properties. It should be clear that this chapter is a natural extension for the case of LTI systems that may, or may not, be BIBO stable, and for more general classes of inputs than those that can be handled by the Fourier transform (with the exception of some eternal signals where we allowed impulse functions in the expression of the Fourier transform). Moreover, our focus was on inputs that are applied at some initial time, $t_0 = 0$ without loss of generality, and to systems that may not be at rest at t_0. By working in the s-domain using the Laplace transform, we can handle both the ZSR and the ZIR in a common framework. In fact, these two responses have much in common in the sense that they give rise to time functions, or modes, that depend on the poles of the TF of the system, as highlighted by Eq. (6.10).

Given how much information the poles of the TF convey about the stability and time response properties of a system, changing the location of these poles emerges as a viable strategy for altering the system behavior to meet certain requirements or specifications such as: stability, damping properties in step response, and so forth. Feedback control using a compensator $C(s)$ provides a mechanism for changing the location of the poles and thereby "controlling" the system to achieve a set of design specifications. This is a whole topic in itself, covered in a controls course that builds on the material presented in this book.

Appendix A
Common Signals

A.1 Continuous Signals

Exponential Signals e^{at}, where a is a real number. If $a = 0$, we have a constant function. If $a > 0$, we have a growing exponential. If $a < 0$, we have a decaying exponential. In this case, we refer to $\frac{1}{|a|}$ as the *time constant*. These signals are important because they arise in the solution of *differential equations*.

Sinusoidal Signals $\cos(\omega_0 t + \phi)$ and $\sin(\omega_0 t + \phi)$. These signals are *periodic*, with fundamental period $T_0 = \frac{2\pi}{\omega_0}$. Sine is an *odd* function and cosine is an *even* function. ϕ is a phase offset.

Complex Exponential Signals $e^{j\omega_0 t}$. By Euler's formula,

$$e^{j\omega_0 t} = \cos(\omega_0 t) + j\,\sin(\omega_0 t)$$

and thus we see that $e^{j\omega_0 t}$ is periodic with radian frequency ω_0 and period $T_0 = \frac{2\pi}{\omega_0}$.
From Euler's formula, we have that

$$\cos(\omega_0 t) = \frac{1}{2}e^{j\omega_0 t} + \frac{1}{2}e^{-j\omega_0 t}$$

and thus the *real* signal $\cos(\omega_0 t)$ is the sum of two *complex* signals, one with "positive frequency" ω_0 and one with "negative frequency" $-\omega_0$.
We have a similar result for sin:

$$\sin(\omega_0 t) = \frac{1}{2j}e^{j\omega_0 t} - \frac{1}{2j}e^{-j\omega_0 t}$$

For each t, the complex number $z = e^{j\omega_0 t}$ has magnitude 1 and argument $\omega_0 t$. Thus, the function $e^{j\omega_0 t}$ corresponds to rotating counterclockwise around the unit

© The Author(s), under exclusive license to Springer Nature Switzerland AG 2022
S. Lafortune, *A Guide to Signals and Systems in Continuous Time*,
https://doi.org/10.1007/978-3-030-93027-1

circle on the complex plane with velocity of $\frac{\omega_0}{2\pi}$ rotations per second. Similarly, the function $e^{-j\omega_0 t}$ corresponds to rotating clockwise around the unit circle on the complex plane with velocity of $\frac{\omega_0}{2\pi}$ rotations per second. This is how one should think of negative frequencies when dealing with complex exponentials.

The family of complex signals $e^{jk\omega_0 t}$ is an example of harmonically related signals. These signals are very important as they are used for the Fourier series representation of periodic signals.

Exponential + Sinusoidal Signals $e^{\sigma_0 t}\cos(\omega_0 t)$. These are sinusoidal signals whose "envelopes" are determined by the exponential $e^{\sigma_0 t}$.

Unit Ramp

$$r(t) = \begin{cases} t & \text{if } t \geq 0 \\ 0 & \text{if } t < 0 \end{cases}$$

Sinc Function

$$\text{sinc}(t) = \begin{cases} \frac{\sin \pi t}{\pi t} & \text{if } t \neq 0 \\ 1 & \text{if } t = 0 \end{cases} \tag{A.1}$$

Note that by L'Hôpital's[1] rule, $\lim_{t \to 0} \frac{\sin \pi t}{\pi t} = 1$, which justifies the assignment $\text{sinc}(0) = 1$. This even signal crosses the horizontal axis at $t = 1, 2, \ldots$

Some books define the sinc function without the π in both the numerator and denominator in Eq. (A.1); to avoid confusion, we refer to this case as the *sampling* function $Sa(t) = \frac{\sin t}{t}$ and use exclusively the sinc function as defined in Eq. (A.1) in this book.

A.2   Piecewise-Continuous Signals

Unit Step

$$u(t) = \begin{cases} 1 & \text{if } t > 0 \\ 0 & \text{if } t < 0 \end{cases}$$

In general, it is better to leave the unit step undefined at the point of discontinuity; however, we often set $u(0) = 1$ if we need a value at $t = 0$.

---

[1] Guillaume François Antoine de l'Hôpital was a French mathematician, 1661–1704.

Note that we can write $r(t) = tu(t)$.

Rectangular Pulse

$$\text{rect}(t/T) = \begin{cases} 1 & \text{if } |t| < T/2 \\ 0 & \text{if } |t| > T/2 \end{cases}$$

Notation: T in $\text{rect}(t/T)$ is the width of the pulse and is not a period.

Signum Function

$$\text{sgn}(t) = \begin{cases} 1 & \text{if } t > 0 \\ -1 & \text{if } t < 0 \end{cases}$$

Appendix B
Proofs

The proofs in this appendix are "standard" in the sense that they follow similar steps to, and are at a comparable level of mathematical rigor as, the proofs of the corresponding results in most undergraduate engineering signals and systems textbooks. We have included these proofs for the sake of making this book as self-contained as possible. In many proofs, the order of integration and/or of summation is changed; it is implicitly assumed that sufficient conditions for these operations hold whenever they are performed.

B.1 Proofs of Main Results

B.1.1 Convolution Integral Theorem

Proof

Step 1 Recall the $p_n(t)$ functions that were considered when defining the unit impulse, but change the notation by replacing n by $\frac{1}{\epsilon}$, i.e.,

$$p_\epsilon(t) = \begin{cases} \frac{1}{\epsilon} & \text{if } |t| \leq \frac{\epsilon}{2} \\ 0 & \text{otherwise} \end{cases}$$

Then, $\lim_{\epsilon \to 0} p_\epsilon(t) = \delta(t)$.

Next, we use the $p_\epsilon(t)$ functions to do a *staircase approximation* of the input $x(t)$. Namely, write

$$x(t) \approx \sum_{k=-\infty}^{\infty} x(k\epsilon)\epsilon p_\epsilon(t - k\epsilon) \tag{B.1}$$

with the approximation becoming exact as $\epsilon \to 0$.

Step 2 Define $h_\epsilon(t)$ to be the output (ZSR) of the system due to input $x(t) = p_\epsilon(t)$. Since $\lim_{\epsilon \to 0} p_\epsilon(t) = \delta(t)$, we conclude that

$$\lim_{\epsilon \to 0} h_\epsilon(t) = h(t)$$

where $h(t)$ is the impulse response of the system.

By *linearity* and *time-invariance*, we get from Eq. (B.1) above that

$$y(t) \approx \sum_{k=-\infty}^{\infty} x(k\epsilon)\epsilon h_\epsilon(t - k\epsilon) \tag{B.2}$$

with the approximation again becoming exact as $\epsilon \to 0$.

Step 3 Finally, look at Eq. (B.2) *for each fixed t* and let ϵ go to 0. As $\epsilon \to 0$, the RHS of Eq. (B.2) becomes the *integral*

$$y(t) = \int_{-\infty}^{\infty} x(\tau)[\lim_{\epsilon \to 0} h_\epsilon(t - \tau)]d\tau = \int_{-\infty}^{\infty} x(\tau)h(t - \tau)d\tau$$

To see why the RHS of Eq. (B.2) becomes an integral as $\epsilon \to 0$, recall the definition of the integral and observe that, in the limit, $k\epsilon$ becomes the continuous variable τ and ϵ becomes the differential $d\tau$. Alternatively, observe that Eq. (B.2) approximates the area under the function (of independent variable τ) $x(\tau)h(t - \tau)$. Thus, as $\epsilon \to 0$, the summation on the RHS of Eq. (B.2) is indeed the integral of the function $x(\tau)h(t - \tau)$, i.e., the *convolution integral*.   Q.E.D.

B.1.2   BIBO Stability and IR

Proof

$$|y(t)| = \left| \int_{-\infty}^{\infty} h(\tau)x(t - \tau)d\tau \right| \leq \int_{-\infty}^{\infty} |h(\tau)||x(t - \tau)|d\tau \leq M_x \int_{-\infty}^{\infty} |h(\tau)|d\tau$$

where we have invoked the fact that the input is bounded by M_x. Hence, the output is also bounded when the IR is absolutely integrable. This shows the *sufficiency* of the condition.

We prove *necessity* (i.e., BIBO stability implies absolute integrability of the IR) by *contrapositive*.[1] Suppose that the IR is not absolutely integrable, i.e.,

$$\int_{-\infty}^{\infty} |h(\tau)|d\tau \text{ " } = \text{ " } \infty$$

---

[1] We follow the proof strategy in [9].

Consider the convolution integral for some *fixed time* $t = t_1$:

$$y(t_1) = \int_{-\infty}^{\infty} h(\tau)x(t_1 - \tau)d\tau$$

Consider the function $x(\tau)$ in the convolution integral and pick the *bounded input*

$$x(\tau) = \text{sgn}[h(t_1 - \tau)]$$

which means that $x(t_1 - \tau) = \text{sgn}[h(\tau)]$. Substituting this bounded input into the convolution integral yields

$$y(t_1) = \int_{-\infty}^{\infty} h(\tau)\text{sgn}[h(\tau)]d\tau = \int_{-\infty}^{\infty} |h(\tau)|d\tau \text{ “} = \text{” } \infty$$

since the IR is assumed not to be absolutely integrable. Hence, we have identified a bounded input that produces an unbounded output, which means that BIBO stability does not hold. This completes the proof of the necessity direction of the "if and only if."　　　　　　　　　　　　　　　　　　　　　　　　　　　　Q.E.D.

B.1.3   Parseval's Theorem

Proof of Parseval's Theorem for Fourier Series

$$\frac{1}{T_0} \int_{<T_0>} |x(t)|^2 dt = \frac{1}{T_0} \int_{<T_0>} \left[\sum_{k=-\infty}^{\infty} X[k]e^{jk\omega_0 t} \right] \left[\sum_{n=-\infty}^{\infty} X[n]e^{jn\omega_0 t} \right]^* dt$$

$$= \frac{1}{T_0} \int_{<T_0>} \left[\sum_{k=-\infty}^{\infty} \sum_{n=-\infty}^{\infty} X[k]e^{jk\omega_0 t} X^*[n]e^{-jn\omega_0 t} \right] dt$$

$$= \frac{1}{T_0} \sum_{k=-\infty}^{\infty} \sum_{n=-\infty}^{\infty} \left[X[k]X^*[n] \int_{<T_0>} e^{jk\omega_0 t} e^{-jn\omega_0 t} dt \right]$$

$$= \sum_{k=-\infty}^{\infty} X[k]X^*[k]$$

$$= \sum_{k=-\infty}^{\infty} |X[k]|^2$$

where the next to last equality is due to the orthogonality property of complex exponentials stated in Eqs. (3.3) and (3.4).　　　　　　　　　　　　　　　　Q.E.D.

Proof of Parseval's Theorem for Energy Signals

$$\int_{-\infty}^{\infty} |f(t)|^2 dt = \int_{-\infty}^{\infty} f(t) f^*(t) dt$$

$$= \int_{-\infty}^{\infty} f(t) \left[\frac{1}{2\pi} \int_{-\infty}^{\infty} F^*(j\omega) e^{-j\omega t} d\omega \right] dt$$

$$= \frac{1}{2\pi} \int_{-\infty}^{\infty} F^*(j\omega) \left[\int_{-\infty}^{\infty} f(t) e^{-j\omega t} dt \right] d\omega$$

$$= \frac{1}{2\pi} \int_{-\infty}^{\infty} F^*(j\omega) F(j\omega) d\omega$$

$$= \frac{1}{2\pi} \int_{-\infty}^{\infty} |F(j\omega)|^2 d\omega$$

Q.E.D.

B.2   Proofs of Fourier Transform Properties

B.2.1   *Convolution and Modulation*

Proof of Convolution Property

$$\mathscr{F}[x(t) * h(t)] = \int_{-\infty}^{\infty} \left[\int_{-\infty}^{\infty} h(t - \tau) x(\tau) d\tau \right] e^{-j\omega t} dt$$

$$= \int_{-\infty}^{\infty} x(\tau) \left[\int_{-\infty}^{\infty} h(t - \tau) e^{-j\omega t} dt \right] d\tau$$

$$= \int_{-\infty}^{\infty} x(\tau) [e^{-j\omega \tau} H(j\omega)] d\tau$$

$$= H(j\omega) \int_{-\infty}^{\infty} x(\tau) e^{-j\omega \tau} d\tau$$

$$= H(j\omega) X(j\omega)$$

One step in the above proof uses the time shifting property, proved below in Sect. B.2.2.   Q.E.D.

Proof of Modulation Property

$$\mathscr{F}[x(t)m(t)] = \int_{-\infty}^{\infty} x(t) \left[\frac{1}{2\pi} \int_{-\infty}^{\infty} M(j\omega')e^{j\omega't} d\omega' \right] e^{-j\omega t} dt$$

$$= \frac{1}{2\pi} \int_{-\infty}^{\infty} M(j\omega') \left[\int_{-\infty}^{\infty} x(t)e^{-j\omega t} e^{j\omega't} dt \right] d\omega'$$

$$= \frac{1}{2\pi} \int_{-\infty}^{\infty} M(j\omega')[X(j(\omega - \omega'))] d\omega'$$

$$= \frac{1}{2\pi} M(j\omega) * Xj(\omega)$$

Q.E.D.

B.2.2 Other Properties

Proof of Time Shifting Property

$$\mathscr{F}[f(t - t_0)] = \int_{-\infty}^{\infty} f(t - t_0)e^{-j\omega t} dt$$

$$= \int_{-\infty}^{\infty} f(\sigma)e^{-j\omega(\sigma + t_0)} d\sigma$$

$$= e^{-j\omega t_0} F(j\omega)$$

Q.E.D.

Proof of Frequency Shifting Property

$$\mathscr{F}[e^{j\omega_0 t} f(t)] = \int_{-\infty}^{\infty} e^{j\omega_0 t} f(t)e^{-j\omega t} dt$$

$$= \int_{-\infty}^{\infty} f(t)e^{-j(\omega - \omega_0)t} dt$$

$$= F(j(\omega - \omega_0))$$

Q.E.D.

Proof of Time Differentiation Property

$$\int_{-\infty}^{\infty} f'(t)e^{-j\omega t} dt = f(t)e^{-j\omega t}|_{-\infty}^{\infty} + j\omega \int_{-\infty}^{\infty} f(t)e^{-j\omega t} dt$$

Since $f(t)$ is assumed to be absolutely integrable, then $f(t) \to 0$ as $t \to \pm\infty$, and thus the first term on the RHS evaluates to zero. The second term is simply $j\omega F(j\omega)$, which completes the proof.   Q.E.D.

Proof of Time Integration Property From the preceding result about time differentiation,

$$\mathscr{F}[f'(t)] = j\omega F(j\omega) \quad \Rightarrow \quad F(j\omega) = \frac{1}{j\omega}\mathscr{F}[f'(t)]$$

Since $f(t) = \int_{-\infty}^{t} f'(\tau)d\tau$, the proof is complete.   Q.E.D.

Proof of Time Scaling Property For $\alpha > 0$,

$$\mathscr{F}[f(\alpha t)] = \int_{-\infty}^{\infty} f(\alpha t)e^{-j\omega t}dt = \frac{1}{\alpha}\int_{-\infty}^{\infty} f(\sigma)e^{-j\omega\sigma/\alpha}d\sigma = \frac{1}{\alpha}F\left(\frac{\omega}{\alpha}\right)$$

The proof is similar for $\alpha < 0$.   Q.E.D.

Proof of Duality Property

$$f(t) = \frac{1}{2\pi}\int_{-\infty}^{\infty} F(\omega)e^{j\omega t}d\omega$$

$$2\pi f(-t) = \int_{-\infty}^{\infty} F(\omega)e^{-j\omega t}d\omega$$

$$2\pi f(-\omega) = \int_{-\infty}^{\infty} F(t)e^{-jt\omega}dt = \mathscr{F}[F(t)]$$

where the last line is obtained by interchanging t with ω.   Q.E.D.

B.3   Proofs of Laplace Transform Properties

B.3.1   Convolution and Time Differentiation

Proof of Convolution Property of Laplace Transform We prove the convolution property for the unilateral Laplace transform first.

$$Y(s) = \mathscr{L}[y(t)] = \mathscr{L}[x(t) * h(t)] = \mathscr{L}[x(t)u(t) * h(t)]$$

$$= \int_{0^-}^{\infty}\left[\int_{0^-}^{\infty} h(t - \tau)x(\tau)d\tau\right]e^{-st}dt$$

$$= \int_{0^-}^{\infty} x(\tau)\left[\int_{0^-}^{\infty} h(t - \tau)e^{-st}dt\right]d\tau$$

$$= \int_{0^-}^{\infty} x(\tau) \left[\int_{-\tau}^{\infty} h(\sigma) e^{-s\sigma} d\sigma \right] e^{-s\tau} d\tau$$

$$= \int_{0^-}^{\infty} x(\tau) \left[\int_{-\tau}^{\infty} h(\sigma) u(\sigma) e^{-s\sigma} d\sigma \right] e^{-s\tau} d\tau$$

$$= \int_{0^-}^{\infty} x(\tau) \left[\int_{0^-}^{\infty} h(\sigma) e^{-s\sigma} d\sigma \right] e^{-s\tau} d\tau$$

$$= \left[\int_{0^-}^{\infty} x(\tau) e^{-s\tau} d\tau \right] \left[\int_{0^-}^{\infty} h(\sigma) e^{-s\sigma} d\sigma \right] = X(s) H(s)$$

Note that we have invoked causality of $h(t)$ and the fact that $x(t)$ is right-sided.

To prove the convolution property for the bilateral Laplace transform, it suffices to replace 0^- by $-\infty$ (and place the subscript B wherever needed) in the above; *note that causality need not be invoked in this case.* Q.E.D.

Proof of Time Differentiation Property of Laplace Transform Using integration by parts, we get

$$\int_{0^-}^{\infty} \frac{df(t)}{dt} e^{-st} dt = f(t) e^{-st} \big|_{0^-}^{\infty} + s \int_{0^-}^{\infty} f(t) e^{-st} dt$$

$$= 0 - f(0^-) + s F(s)$$

whenever $s \in \text{ROC}[F(s)]$ since in that case $\lim_{t \to \infty} f(t) e^{-st} = 0$. Q.E.D.

B.3.2 Time Integration and Time Shifting

Proof of Time Integration Property Let $y(t) = \int_{0^-}^{t} f(\tau) d\tau$. Then,

$$Y(s) = \mathscr{L}[y(t)] = \int_{0^-}^{\infty} y(t) e^{-st} dt$$

$$= y(t) \frac{(-1)}{s} e^{-st} \big|_{0^-}^{\infty} + \frac{1}{s} \int_{0^-}^{\infty} f(t) e^{-st} dt$$

$$= 0 + \frac{y(0^-)}{s} + \frac{1}{s} F(s)$$

$$= \frac{1}{s} F(s)$$

where the next to last equality is due to the fact that we take $s \in \text{ROC}(F(s))$ and the last equality is due to the fact that $y(0^-) = 0$ by definition. Q.E.D.

Proof of Time Shifting Property The property is proved by a change of the variable
of integration:

$$\mathscr{L}[f(t-t_0)u(t-t_0)] = \int_{0^-}^{\infty} f(t-t_0)u(t-t_0)e^{-st}dt$$

$$= \int_{t_0^-}^{\infty} f(t-t_0)e^{-st}dt$$

$$= \int_{0^-}^{\infty} f(\sigma)e^{-s(\sigma+t_0)}d\sigma$$

$$= e^{-st_0}F(s)$$

Q.E.D.

B.3.3   Final Value Theorem

Proof of Final Value Theorem

$$sF(s) - f(0^-) = \int_{0^-}^{\infty} f'(t)e^{-st}dt$$

$$\lim_{s\to 0}[sF(s) - f(0^-)] = \int_{0^-}^{\infty} f'(t)dt$$

$$= \lim_{t\to\infty} f(t) - f(0^-)$$

from which the result follows.   Q.E.D.

References

1. Haykin S, Van Veen B (2003) Signals and systems, 2nd edn. Wiley, New York
2. Kamen EW, Heck BS (2007) Fundamentals of signals and systems, 3rd edn. Pearson, Upper Saddle River
3. Kudeki E, Munson DC Jr (2009) Analog signals and systems. Pearson, Upper Saddle River
4. Lathi BP, Green R (2021) Signal processing and linear systems, 2nd edn. Oxford University Press, Oxford
5. Lee EA, Varaiya P (2003) Structure and interpretation of signals and systems. Addison-Wesley, Reading
6. Oppenheim AV, Willsky AS, Young IT (1983) Signals and systems. Prentice-Hall, Englewood Cliffs
7. Phillips CL, Parr JM (1995) Signals, systems and transforms. Prentice-Hall, Englewood Cliffs
8. Senior TAB (1986) Mathematical methods in electrical engineering. Cambridge University Press, Cambridge
9. Soliman SS, Srinath MD (1998) Continuous and discrete signals and systems, 2nd edn. Prentice-Hall, Englewood Cliffs
10. Ulaby FT, Yagle AE (2016) Engineering signals and systems in continuous and discrete time, 2nd edn. National Technology and Science Press, Allendale

Index

A
Aliasing, 67
Analog to digital converter, 69
Anti-aliasing filter, 67

B
Band-limited signal, 37, 59
Bandwidth, absolute, 62
Bandwidth, RMS, 63
Bandwidth, 3dB, 63
Baseband, 59
Bode plot, 49
Butterworth filter, 61

C
Causal system, 12, 25
Complex exponential function, 99
Complex impedances, 32
Conjugate symmetry, 30, 37, 48
Convolution integral theorem, 20
Convolution property, 52, 75
Convolution, table, 23
Critical damping, 93

D
Damping ratio, 34, 93
Demodulation, synchronous, 59
Differentiator, 54
Digital to analog converter, 70
Double-sideband suppressed-carrier amplitude
    modulation, 59
Duality, 56
Dynamic system, 12

E
Eigenfunction, 26
Eigenfunction theorem, 26
Energy spectrum, 56

F
Feedback connection, 76, 93
Filetring, 57
Filters, ideal, 58
Filters, realizable, 61
Final value theorem, 92
Fourier series and LTI systems, 42
Fourier series, combined trigonometric, 37
Fourier series, existence conditions, 36
Fourier series, exponential, 35
Fourier series, trigonometric, 37
Fourier transform, definition, 45
Fourier transform, existence conditions, 48
Fourier transform, table, 49
Frequency differentiation property, 55, 84
Frequency-division multiplexing, 60
Frequency response function, 28, 45
Frequency shifting property, 54, 85
Frequency spectrum, 37, 48

G
Graphical convolution, 24

I
Impulse response, 20
Integrator, 21
Interpolation formula, 66

© The Author(s), under exclusive license to Springer Nature Switzerland AG 2022
S. Lafortune, *A Guide to Signals and Systems in Continuous Time*,
https://doi.org/10.1007/978-3-030-93027-1

L

Laplace transform, definition, 74
Laplace transform, existence conditions, 77
Laplace transform, inverse, 79
Laplace transform, table, 78
Linearity of systems, 15

M

Memoryless system, 11
Mixing, 47
Modes of system, 88
Modulation, 47, 59
Modulation property, Fourier transform, 46

N

Natural frequency, 34, 93
Negative frequencies, 100

O

Overdamping, 93
Overshoot, frequency domain, 34

P

Parallel connection, 3, 23, 75
Parseval's theorem, Fourier series, 40
Parseval's theorem, Fourier transform, 56
Partial fraction expansion, 80
Partial fraction expansion, complex roots, 83
Partial fraction expansion, distinct roots, 81
Partial fraction expansion, repeated roots, 82
PID compensator, 95, 96
Plant, 94
Poles, 81, 91
Power spectrum, 41
Problem IO, 3, 22, 43, 53, 75
Problem RE, 3, 43, 53, 75
Problem SI, 3, 43, 53, 75
Proportional feedback, 94

R

Region of convergence, 77
Residue, 81
Resonance, 34
Root locus, 94
Routh-Hurwitz criterion, 91, 96

S

Sampling frequency, 65
Sampling theorem, 65
Second-order system, 34, 92
Series connection, 3, 22, 75
Sifting property of unit impulse, 18
Signal, 1
Signal, analog, 2
Signal, discrete, 2
Signals, energy, 7
Signals, periodic, 6
Signals, power, 8
Signum function, 50, 101
Sinc function, 100
Sine-in sine-out law, 31, 42
Sinusoidal function, 99
Stability, asymptotic, 15, 91
Stability, BIBO, 14, 25, 91
State of dynamic system, 12
Static system, 11, 25
Step response, 22
System, 2

T

Time differentiation property, 54, 84
Time integration property, 54, 84
Time-invariance of systems, 15
Time reversal, 8
Time scaling, 8, 19
Time scaling property, 55, 85
Time shifting, 8
Time shifting property, 40, 54, 85
Train of impulses, 51, 67
Transfer function, 28, 73, 86

U

Underdamping, 93
Unit impulse signal, 17

W

Whittaker-Shannon interpolation formula, 66

Z

Zero-input response, 13, 87
Zero-order hold, 69
Zeros, 81
Zero-state response, 13, 86

Printed in the United States
by Baker & Taylor Publisher Services